ELECTROSTATICS

Edited by **Hüseyin Canbolat**

Electrostatics
Edited by Hüseyin Canbolat

Published by InTech
Janeza Trdine 9, 51000 Rijeka, Croatia

Publishing Process Manager Martina Durovic

Technical Editor Teodora Smiljanic

Cover Designer InTech Design Team

Additional hard copies can be obtained from orders@intechopen.com

Electrostatics, Edited by Hüseyin Canbolat
p. cm.
ISBN 978-953-51-0239-7

For EU product safety concerns:
IN TECH d.o.o., Prolaz Marije Krucifikse Kozulić 3, 51000 Rijeka, Croatia,
info@intechopen.com or visit our website at intechopen.com.

Meet the editor

Dr Hüseyin Canbolat was born in Adana in 1966. He received his BS and MS degrees from the Middle East Technical University, Ankara, Turkey, in 1989 and 1993 respectively, and a PhD degree in Electrical Engineering from Clemson University, Clemson, SC, USA, in 1997. As an Associate Professor he joined the Department of Electrical Engineering at Mersin University in 1998 and at the beginning of 2012 he joined the Department of Electronics and Communication Engineering at Yildirim Beyazit University, Ankara, Turkey. His research interests include control systems with applications to robotic and mechatronic systems, microelectromechanical systems (MEMS) and renewable energy systems. He is a member of IEEE and the Chamber of Electrical Engineers of Turkey.

Contents

Preface

Electrostatics is the branch of electricity dealing with the phenomena created by stationary charges. Electric charges exert force on each other according to the Coulomb's law.

Electrostatic phenomena include many examples. Some simple examples are everyday experiences of attracting the plastic wrap to a hand after it is removed from a package and small electroshocks during the first contact with a metal object, especially after working with an electrical appliance. However, there are more serious phenomena caused by electrostatic charges, such as the apparently spontaneous explosion of grain silos or damage to electronic components during manufacturing. It is also used in some modern day technology, such as operation of photocopiers.

Electrostatics involves the buildup of charge on the surface of objects due to contact with other surfaces. Although charge exchange happens whenever any two surfaces contact and separate, the effects of charge exchange are usually only noticed when at least one of the surfaces has a high resistance to electrical flow. This is because the charges that transfer to or from the highly resistive surface are more or less trapped there for a long enough time for their effects to be observed. These charges then remain on the object until they either bleed off to the ground or are quickly neutralized by a discharge.

In this book, the authors provide state-of-the-art research studies on electrostatic principles or include the electrostatic phenomena as an important factor. The chapters cover diverse subjects, such as biotechnology, bioengineering, actuation of MEMS, measurement and nanoelectronics. Hopefully, the interested readers will benefit from the book in their studies. It is probable that the presented studies will lead the researchers to develop new ideas to conduct their research.

Dr Hüseyin Canbolat
Yildirim Beyazit University, Ankara,
Turkey

Part 1

Electrostatics in Biological Sciences

1

Electrostatic Interactions in Dense DNA Phases and Protein-DNA Complexes

A. G. Cherstvy
Institute of Complex Systems, ICS-2, Forschungszentrum Jülich, Jülich,
Institute for Physics and Astronomy, University of Potsdam, Potsdam-Golm,
Germany

1. Introduction[1]

Charges. Many constituents of living cells bear large charges on their surfaces. The list includes DNA/RNA nucleic acids [1], cellular lipid membranes [2], DNA-binding [3,4] and architectural [5,6] proteins, natural ion channels [7] and pores, elements of cytoskeleton networks [8], and molecular motors. ES interactions on the nano-scale often dominate the physical forces acting between these components in the last 1-3 nm prior to surface-surface contact, often governing their spontaneous assembly and long-range spatial ordering. There has been a number of excellent reviews covering the general principles of ES effects in nucleic acids [9,10], proteins [11,12,13], lipid membranes [2,14,15], and some other bio-soft-matter systems [16]. Salt- and pH-sensitivity of ES forces provides cells with a useful handle to direct/tune the pathways of many biological processes. Among them are DNA-DNA, protein-DNA [17] and protein-protein ES interactions [18], DNA compactification into higher-order structures [19,20], DNA spooling inside viral shells [21], actin aggregation, RNA folding [10,22], and ion translocation through membrane pores. ES forces modulate structure and control functioning of sub-cellular supra-molecular assemblies [23,24] and can affect cell-cell interactions in tissues [25].

Over the last years, the ES mechanisms of some DNA-related phenomena mentioned above have been developed in our group. General concepts of the PB theory often give a satisfactory physical description of ES properties of molecules in solution and macromolecular complexes. Below, we try to keep the presentation on illustrative level avoiding complicated algebra: all analycal expressions, the details of their derivation, and regimes of applicability can be found in the original papers cited. We rather focus on underlying physical mechanisms, comparing the system behavior under varying conditions. We often treat ES forces in dense, weakly fluctuating structures/complexes, where entropic effects are weak and can be neglected. Because of limited space, we focus on latest ES-motivated developments from other groups, trying to position our research in this context.

[1]Abbreviations: ES, electrostatic; HB, hydrogen bond; PE, polyelectrolyte; PB, Poisson-Boltzmann; PEG, polyethylene glycol; DH, Debye-Hückel; EM, electron microscopy; AFM, atomic force microscopy; bp, base pair; kbp, kilo base pair; [DNA], DNA concentration; [salt], salt concentration; ds, double stranded; ss, single stranded; CL, cationic lipids; LC, liquid-crystalline; GNP, gold nano-particle; hom, homologous; NCP, nucleosome core particle; PDB, Protein Data Bank.

Outline. The main aim of this chapter is to provide a review of recent advances in the theory of ES interactions in dense assemblies of DNAs and to discuss some ES aspects of protein-DNA recognition and binding. These subjects have been the main area of my scientific activity in the last several years. ES effects on different levels of DNA organization *in vivo* and *in vitro* are considered below. We overview e.g. the biophysical principles behind DNA-DNA ES interactions, DNA complexation with CL-membranes, DNA condensates, DNA cholesteric phases and touch on DNA spooling inside viruses. For DNA-protein complexes, the effects include ES recognition and binding. For these systems, we develop theoretical frameworks and computational approaches to describe physical-chemical mechanisms of structure formation that allow us later to anticipate some biological consequences.

First, we focus on theoretical concepts used in derivation of the ES interaction potential of two parallel double-helical DNAs immersed in electrolyte solution [9]. The linear PB theory developed for this system [26] accounts for a low-dielectric DNA interior and spiral distribution of negative phosphate charges on DNA periphery. We discuss the regimes of applicability of this linear theory, in application to interaction of DNAs partly neutralized by adsorbed counterions. This theory and its modifications have allowed us to rationalize a number of experimental observations regarding the behavior of DNAs in columnar hexagonal phases (Sec. 2), dense cholesteric DNA assemblies (Sec. 5), the decay length of DNA-DNA ES repulsion in mono-valent salts, the region of DNA-DNA attraction in the presence of multivalent cations (Sec. 2), as well as DNA condensation into toroids (Sec. 3). A separate domain of our research deals with interaction-induced adjustment of DNA helical structure, DNA-DNA sequence recognition and pairing (Sec. 7), as well as DNA-DNA friction (Sec. 8). We also overview DNA melting and hybridization in dense DNA lattices, Sec. 9.

In the second part, we focus on ES recognition between DNA and DNA-binding proteins in their complexes. We propose a model of DNA sequence recognition by relatively small proteins (e.g., transcription factors) based on complementarity of charge patterns on DNA target site and bound protein (Sec. 10). For relatively large proteins, we support the theoretical conclusions by a detailed bioinformatic statistical analysis of charge patterns along interfaces of various protein-DNA complexes, as extracted from their PDB entries. We decipher the reasons why large structural protein-DNA complexes of pro- and eu-karyotic organisms do involve a substantial ES component in their recognition (Sec. 10). On the contrary, DNA recognition by small DNA-binding proteins appears to be ES-non-specific, being likely governed by HB formation.

Every section below starts with a short introduction to the subject, followed by a presentation of basic theoretical concepts and discussion of main results, and it ends with some perspectives for future developments and possible model improvements. The content of this chapter is based on the recent perspective article [27].

2. ES forces between DNA duplexes

Counterion condensation. B-DNA is one of the most highly charged bio-helices, with one elementary charge e_0 per $\approx$nm^2 on the surface at standard pH and physiological [salt]. These charges are the phosphate negative groups located on DNA periphery, forming a duplex with 10-10.5 bp per helical turn of $H\approx 34$Å and non-hydrated DNA radius of $a \approx 9$ Å. More that ~75% of DNA charge is neutralized by counterions adsorbed onto it from solution. The

Manning theory [28] predicts θ_M=76% of charge compensation for mono-valent (z=1) and θ_M=92% for tri-valent (z=3) cations. In the DNA model as a thin long linear PE at vanishing [salt], the neutralization fraction is predicted to be

$$\theta_M = 1 - 1/(z\xi), \tag{1}$$

where ξ is the ratio of the Bjerrum length ($l_B \approx 7.1$ Å in water) to the axial PE inter-charge separation, $b \approx 1.7$ Å for the bare DNA. Recent experiments on DNA translocation through nm-sized solid-state nano-pores enabled measuring the compensation fractions θ [29,30], often in good agreement with the Manning theory.

Cation binding. DNA structure offers well-defined sites for counterion binding. Depending on chemical nature and valence, cations bind in DNA grooves, on DNA strands, or both. The distribution and binding equilibrium of adsorbed cations result in a distinct pattern of charges on DNA surface that, in turn, dictates the properties of DNA-DNA ES forces. It also affects intrinsic DNA helical structure and conformation DNA adopts in solution [31]. ES forces are believed to dominate the interaction of parallel DNAs in the last 20Å prior to surface-surface contact, because of still relatively large residual DNA charge density after condensation of counterions.

DNA-DNA hydration force created by overlapping patterns of structured water molecules on DNA surfaces is another alternative [32]. Close similarities in the magnitude and decay length of repulsive forces in the last 1-2 nm prior to the contact measured by osmotic stress technique in simple-salt solutions for DNA, some net-neutral polymers [33,34] and lipid membranes [14,35] favor the hydration force picture. Extreme sensitivity of DNA-DNA forces measured to the chemical nature and valence of cations added, not expected to affect strongly the close-range hydration forces, favors however the ES mechanism of DNA-DNA force generation. In particular, DNA-DNA attraction in the presence of multi-valent cations can be rationalized by our ES models, see below.

The pattern of condensed cations bears some/strong correlations to the helical symmetry of DNA phosphates, forming a "lattice" of alternating positive-negative charges along the DNA axis, Fig. 1. ES forces between these periodic arrays of charges might turn from repulsion to attraction for well-neutralized DNAs. Attractive DNA-DNA forces have been systematically measured by the osmotic stress technique in dense columnar hexagonal DNA assemblies in the presence of some di- and many tri-valent cations at ≈1 nm between the surfaces [36,37], Fig. 2, while purely repulsive forces have been detected with mono-valent salts [38]. The list of DNA condensing agents includes multi-valent cations (cohex^{3+}, spermine^{4+}, spermidine^{3+}), some highly positively charged proteins and polypeptides (poly-Lys and poly-Arg, protamines, H1 histones), as well as concentrated solutions of neutral PEG polymers. The latter are excluded from the DNA phase, exerting an external osmotic pressure onto the DNA lattice. Some ions from this list interact with DNA in natural environments, such as spermidine^{3+} present in many bacteria in 1-3 mM concentrations [39], protamines that are abundant in sperm heads, as well as putrescine^{2+} and spermidine^{3+} vital for DNA compaction in some T-even bacterio-phages [40].

Duplex-duplex ES forces. A number of theoretical models have been developed in the last two decades to provide a physical rationale for DNA-DNA attraction, including some recent advances [41,42]. In one group of models, the spatio-temporal correlations of cations stem from the inherent DNA structure, which render DNA-DNA attraction possible via a "zipper effect". In other models, beyond the PB limit, the *correlated* fluctuations in the density profiles of condensed cations give rise to attraction [43,44,45], even for DNAs modeled as a uniformly charged PE rods. The period of oscillatory charge density waves on PE surfaces in these models is largely decoupled from intrinsic DNA charge periodicity. To save space, we address the reader to a comprehensive review [9] focused primarily on ES DNA-DNA forces. It provides a broad coverage, physical comparison, and analysis of applicability regimes for various models of PE like-charge attraction. In this chapter, we target primarily *new developments* in the theory of DNA-DNA and DNA-protein ES interactions. DNA-DNA attraction has also been extensively investigated by computer simulations [46,47,48,49], for diverse models for DNA structure, the shape and binding specificity of counterions, as well as for various solvent models implemented.

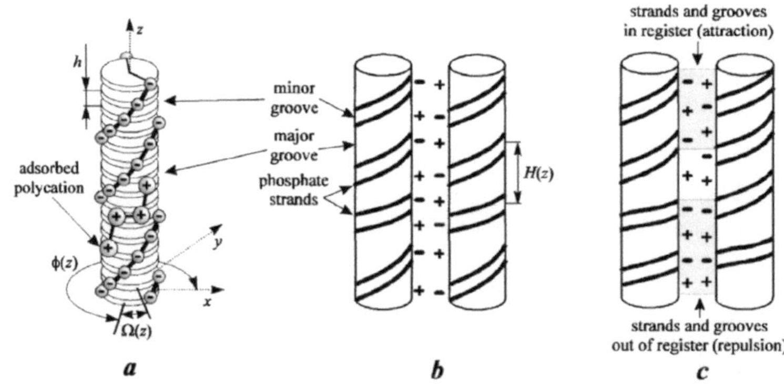

Fig. 1. Schematics of cation-decorated DNA duplexes (a) and interacting hom vs. non-hom sequences (b, c). Positive-negative charge zipper motif that ensures DNA-DNA ES attraction is shown in part (b). The image is reprinted from Ref. [53], subject to APS-2001 Copyright.

The helicity of DNA charges renders the ES potential close to the double helix helically symmetric. When two DNAs approach one another in electrolyte, these helical potential profiles overlap. This affects non-trivially DNA-DNA ES forces, on top of ES repulsion of uniformly charged rods. The exact theory of ES forces between two long parallel double-helical macromolecules was developed in 1997 by A. Kornyshev and S. Leikin [26]. This elegant linear PB theory explicitly accounts for the DNA charge helicity and its low-dielectric hydrophobic core (permittivity of $\varepsilon_c \approx 2$).

The model implies two distinct populations of cations around the DNA. The first one is the Manning's fraction of cations is strongly/irreversibly adsorbed in DNA grooves/strands, while the remaining DNA charge is shielded by electrolyte ions in DH linear manner. DNA ES potential renormalized in this fashion often does not exceed 25 mV, rendering the linear PB model applicable to description of interacting clouds of mobile ions around two partly-neutralized DNAs. Both DNA phosphates and condensed cations in the middle of DNA grooves are modeled below as thin continuous helical lines of charges. Thermal smearing of charge pattern can be incorporated via the Debye-Waller factor [9] that reduces the magnitude of the helical harmonics $a_{1,2}$, see below.

Attraction vs. repulsion. The theory predicts ES attraction of well-neutralized DNAs with the majority of cations adsorbed in the major groove, as pioneered in Ref. [50]. This arrangement of charges facilitates a periodic positive-negative charge alternation along the DNA axis. In physical terms, a DNA-DNA attraction emerges from a *zipper-like ES matching* of phosphate groups of one DNA with the cations adsorbed in a regular fashion in the grooves of another DNA. Many large or extended multivalent DNA-condensing cations are indeed known to bind preferentially into the major DNA groove, preferred both from interaction and steric point of view. Correlated ES potential alternations thus generate charge interlocking along the DNA-DNA contact and give rise to DNA-DNA ES attraction, see below. The mathematical apparatus used for deriving DNA-DNA forces, ES and chemical features of counterion binding, as well as applicability regimes of this mean-filed continuum DH-Bjerrum PB theory are discussed in details in excellent review [9].

Further developments of this theory enabled us to incorporate fine realistic details of DNA structure, such as a discrete nature of adsorbed cations [51] and sequence-specific pattern of the twist angles [52] between the adjacent DNA bps [53]. The models for description of interaction- and T-mediated rearrangements of condensed cations on DNA surfaces [54], torsional flexibility of DNA backbone [55], some soliton-like DNA twist "defects" [56], and DNA helical "straightening" in dense phases [57] have also been developed. ES forces between non-parallel infinitely long [58] and finite-length [59] DNAs were computed and the detailed statistical theory of dense DNA assemblies has been worked out [60].

Basic equations. A number of outcomes of this theory are in quantitative agreement with a number of exprimental observations available for DNA assemblies. These include the decay length of DNA-DNA repulsion in simple salt solutions and attraction at R=28-32Å in the presence of multivalent cations, Fig. 2. Also, DNA azimuthal frustrations [55,61], DNA straightening [62], and a reduced positional order observed in dense DNA lattices [63] have been rationalized. Recent developments unraveled the effects of DNA thermal undulations [64,65] and have shown that duplex-duplex ES forces might get amplified in DNA columnar phases at finite T, as compared to T=0 case. Recently, the implications of binding equilibrium of finite-size ions on DNA-DNA ES forces have been clarified [66]. A number of biological consequences of computed ES duplex-duplex forces were analyzed in excellent recent perspective [67].

DNA-DNA ES interaction energy in electrolyte solution can be approximated as the sum of the first helical interaction harmonics a_n [9]

$$E(R,L) \approx L\left[a_0(R) - a_1(R)\cos\delta\phi + a_2(R)\cos 2\delta\phi\right]. \tag{2}$$

These positive coefficients decay nearly exponentially with DNA-DNA separation R, Fig. 3, and their values depend on partitioning of cations on DNA and DNA charge compensation θ as follows [50]

$$a_0(R) = \frac{8\pi^2\bar{\sigma}^2 a^2}{\varepsilon}\left[\frac{(1-\theta)^2 K_0(\kappa_D R)}{\left[\kappa a K_1(\kappa a)\right]^2} - \sum_{n,m=-\infty}^{\infty} \frac{\tilde{f}(n,\theta,f)^2 K_{n-m}{}^2(\kappa_n R) I_m{}'(\kappa_n a)}{\left[\kappa_n a K_n{}'(\kappa_n a)\right]^2 K_m{}'(\kappa_n a)}\right],$$

$$a_{m=1,2}(R) = \frac{16\pi^2\bar{\sigma}^2 a^2}{\varepsilon}\frac{\tilde{f}(m,\theta,f)^2 K_0(\kappa_m R)}{\left[\kappa_m a K_m{}'(\kappa_m a)\right]^2}. \tag{3}$$

Here, the first term in a_0 describes the ES repulsion between uniformly charged "DNA rods", that dominates at large *R*. The second term in a_0 is the image-charge repulsion between the charges on one DNA from image charges (of the same sign) created in a low-dielectric core of another DNA. The duplex-specific DNA-DNA forces are described by $a_{1,2} > 0$ amplitudes. For ideally helical DNAs, the interaction energy scales linearly with the DNA length *L*, while for randomly-sequenced non-ideal DNA fragments a more intricate dependence arises, see Sec. 7. With the cations adsorbed prevalently in the major groove and at large θ values, the a_1-term responsible for ES helix-helix attraction grows. Many DNA-condensing multivalent cations are indeed known to adsorb into the major DNA groove.

In these expressions, parameter *f* controls the partitioning of cations on DNA (at *f*=0 all cations occupy the major groove), $\tilde{f}(n,\theta,f) = f\theta + (-1)^n (1-f)\theta - \cos(n\tilde{\phi}_s)$, $\tilde{\phi}_s \approx 0.4\pi$ is the azimuthal half-width of DNA minor groove, $\bar{\sigma}$ is the surface charge density of DNA phosphates, and $K_n(x)$, $I_n(x)$, $K_n'(x)$, $I_n'(x)$ are the modified Bessel functions of order *n* and their derivatives.

We note that the decay lengths of $a_{n=1,2}$ harmonics, $1/\kappa_n = 1/\sqrt{\kappa^2 + n^2(2\pi/H)^2}$, is a non-trivial function. Not only it contains the DH screening length in 1:1 solution with [salt]= n_0, namely $\lambda_D = 1/\kappa = 1/\sqrt{8\pi l_B n_0}$, but also depends on the DNA helical repeat *H*. We remind here that at physiological conditions $\lambda_D \approx 7-10$ Å, that is n_0 ~0.15-0.1 M of simple salt.

The image-force repulsion is screened with about half as short decay length, compared to the direct charge-charge repulsion. Effectively, the electric field travels a double distance to image charge. This gives rise to a short-range branch of DNA-DNA ES repulsion at *R*<24Å or so (for typical parameters), see Fig. 2. For intermediate *R*=28-32Å, the ES helix-helix attraction overwhelms the short-range image-force repulsion and the direct DH charge repulsion [68]. This renders the net DNA-DNA ES force *attractive* in this range. Typically, at about *R*>35Å the direct DH rod-rod repulsion prevails and DNAs again repel each other.

Note however that the predicted in Fig. 2 short-range repulsion domain is shifted by 3-5Å to smaller DNA-DNA distances, as compared to the measured DNA pressure-distance curves. A possible explanation is that the first, tight hydration shell of DNA, not included in the theory, might effectively increase DNA diameter in experiments and thus prevent direct DNA contacts at *R*≈20Å, shifting the energy curves measured towards larger *R* values.

Another effect is azimuthal frustrations of DNA molecules observed in dense hexagonal DNA lattices [61]. In the theory, they emerge from XY-spin-like $\cos\delta\phi - \cos 2\delta\phi$ dependence of the interaction potential on the mutual DNA rotation angle, $\delta\phi$. Optimization of the interaction energy over all 6 neighboring DNAs on the lattice inevitably "frustrates" the azimuthal order [69]. Frustrated Potts-like states, reminiscent of those for magnetic spin systems, are often preferred for DNA hexagonal lattice in the model [61]. Namely, in the elementary triangle on a lattice, the two differences of the azimuthal DNA angles are

$$\Delta_1 = \pm \arccos\left[1/4 + \sqrt{1 + 2a_1/a_2}/4\right], \tag{4}$$

while the third one is 2 times larger [70].

Future challenges. Below, we overview some challenges for the current theory. One of them is water structuring in the hydration shells around the DNA. Namely, the most interesting features of intermolecular forces, including the attraction region, emerge at DNA densities when the shells of "structured waters" on interacting helices can overlap. Also, a distance-dependent "effective" dielectric constant on the length scale of 1-2 water diameters [11], a modified decay of electric fields close to DNA, a finite diameter and precise geometrical form of DNA-condensing cations (e.g., linear flexible polyamines vs. compact cohex^{3+} ions), as well as a limited applicability of the linear PB model, all these points require more accurate theories to be developed close to DNA surface. The solvation of DNA also requires a microscopic treatment of dielectric environments and polarization states upon counterion binding to the DNA. Not only in the theory, these factors also complicate quantitative predictions of DNA-DNA forces by means of computer simulations. Similar complications in description of ES forces on the nano-scale emerge in modeling of DNA-protein, DNA-membrane, as well as protein-protein complexes (discussed in Sec. 10).

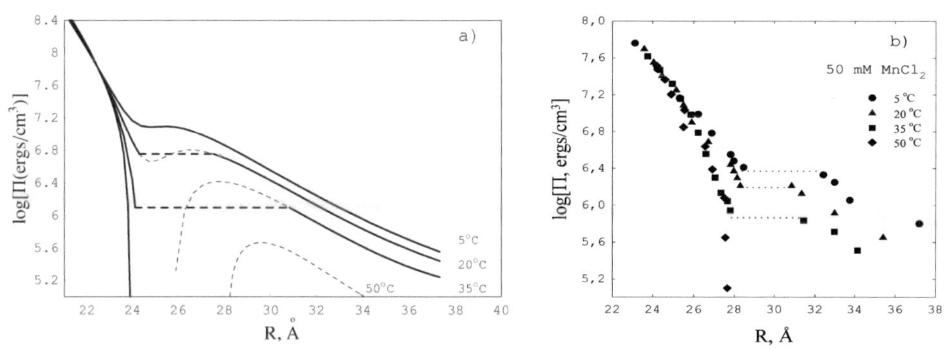

Fig. 2. Theoretically predicted (a) and experimentally measured (b) DNA-DNA forces in dense DNA assemblies at 50 mM $MnCl_2$. The region of DNA-DNA attraction at R=28-32Å detected in experiments corresponds to a spontaneous shrinkage/collapse of the DNA lattice. No azimuthal frustrations on DNA lattice were considered in the model, i.e. $\cos\delta\phi \equiv 0$ for all DNA pairs. Note that, contrary to majority of di-valent cations, Mn^{2+} and Cd^{2+} are capable of generating DNA-DNA attraction under the osmotic stress of PEG [36]. This technique allows to overcome the long-range DH repulsive branch of the potential and thus enhance the helix-mediated DNA-DNA forces shielded with a shorter screening length, $1/\kappa_1$. Parameters: $\theta = 0.85$, $n_0 = 50$ mM. The figure is reprinted from Ref. [54], subject to ACS-2002 Copyright.

3. DNA toroidal condensation

Structure of toroids. One biological manifestation of cation-mediated DNA-DNA attraction is DNA condensation into compact toroidal structures observed in bacteria, viruses, and sperm cells *in vivo* and studied thoroughly *in vitro* [71]. For instance, some bacteria pack their DNAs into robust toroids to protect the genetic material and minimize the frequency of ds-DNA breaks [72]. These radiation-resistant bacteria retain strongly elevated [Mn^{2+}] in their cells to regulate packaging of chromatin fibers, via likely attractive DNA-DNA forces [73]. In mammalian sperm cells, very long DNA is condensed with the help of highly basic Arg-rich proteins protamines into the assembly of interconnected small toroids, as visualized by the AFM technique [74]. DNA compaction inside T5 bacteriophage in the presence of spermine^{4+} also exhibits some toroidal-like arrangements for a part of DNA spool, that is likely to optimize the energetics of DNA packing/encapsidation inside viral shells [75,76].

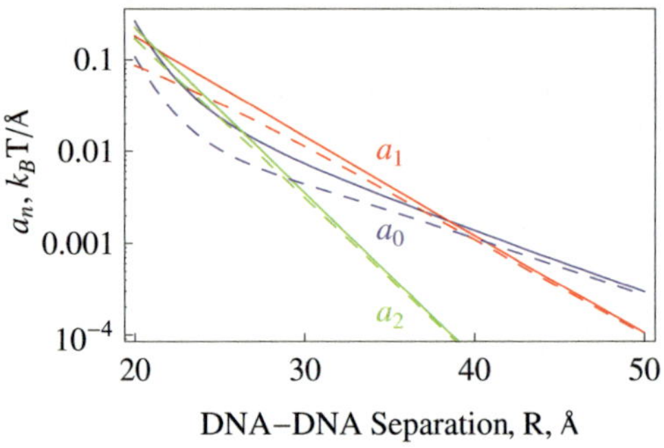

Fig. 3. Dependence of the helical ES harmonics at typical DNA parameters: $\theta = 0.8$, f=0.3, a=9Å, $1/\kappa = 7$ Å. The solid curves are plotted for two DNAs in solution; the results for dense DNA lattices with the Donnan equilibrium are the dashed curves. In the region of DNA-DNA attraction, the first helical a_1 term dominates the interaction energy. The figure is reprinted from Ref. [27], subject to RCS-2011 Copyright.

In vitro, DNA condensates formed in solutions of cohex^{3+}, as visualized by cryo-EM, often reveal a spool-like DNA organization into tori with ~50 nm outer and ~15 nm inner radii [77], with nearly hexagonal local DNA lattice order, Fig. 4. When several DNA chains comprise a torus, the most frequently encountered condensates contain an optimal number of DNA strands. Often, nearly hexagonal toroidal cross-sections are observed, with a completely filled outer DNA shell, which give the most stable aggregates. Such structures maximize the number of attractive DNA-DNA contacts inside the toroid and minimize the number of (relatively unfavourable) DNA contacts with the solvent. It is important to note that DNA-DNA separations in toroids are often R≈28Å, being in the range of DNA-DNA attraction as measured by the osmotic stress technique and as predicted by the theory of DNA-DNA ES interactions, see Fig. 2.

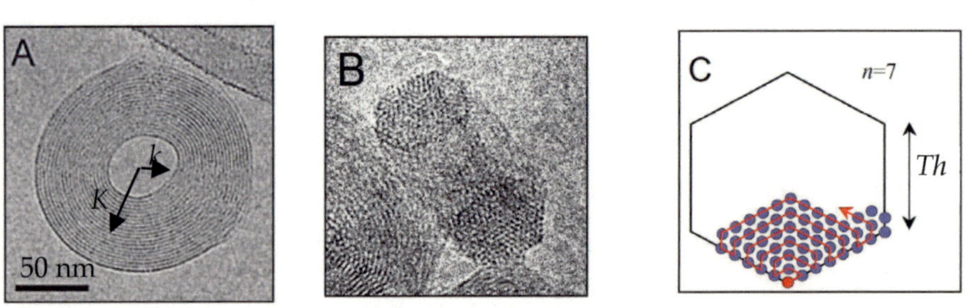

Fig. 4. Cryo-EM images of DNA toroids constructed from 2-3 λ-phage 48.5 kbp long DNAs in 0.2 mM solution of cohex^{3+} (A, B). The mean K and inner k toroidal radii are indicated. One possible model of a defect-free DNA spooling into a torus of generation n=7 in shown in part (C). The image is reprinted from Ref. [80], with permission of IOP.

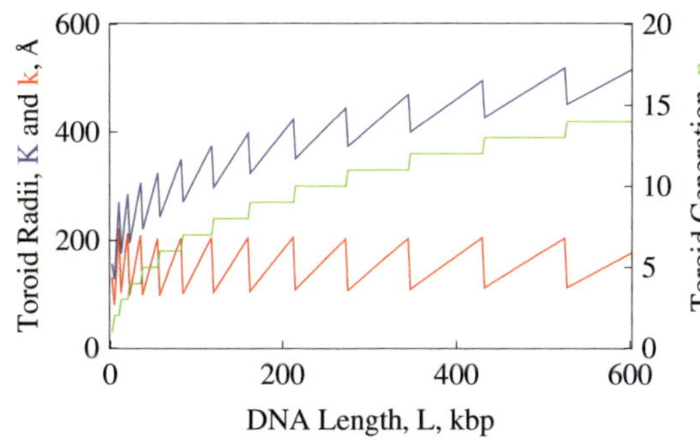

Fig. 5. Radii of DNA toroids of generation *n*, as obtained at relatively strong DNA-DNA attraction of $E_0 = -0.05k_BT$ / Å [80]. The saw-tooth variation of toroid dimensions is due to the growth-by-generation model implemented.

Let us mention one more example of dense DNA assembly, 3D DNA origami structures, where *extremely dense* DNA packing at $R \approx 22$-25Å is realized [78,79]. A successful assembly necessitates ~10-20 mM of $MgCl_2$: the divalent cations are likely to reduce the ES repulsion of DNA strands during the assembly process. The latter is driven by the chemical energy of association of complementary ss-DNA fragments into ds-DNA fragments.

Model and outcomes. Utilizing these facts, we constructed a simple model of DNA toroid growth by generations [80]. Due to a finite value of the DNA bending persistence length lp [81,82], DNA toroids are often preferred over rod-like or (hollow) spherical condensates. During the first stage of compaction, initial DNA circular loop is thermally nucleated and stabilized, with the curvature radius of ~l_p. The growth of DNA toroids is controlled by DNA-DNA attractive ES contacts and by unfavorable energy of DNA elastic deformations.

As the toroidal cross-section increases, the fraction of "missing" DNA-DNA attractive contacts on the toroid periphery progressively decreases (the volume-to-surface ratio grows). This improves the ES attractive energy gain per unit length of DNA compacted, approaching the value one gets for the DNA columnar hexagonal phase, where the pair DNA-DNA interaction is tripled due to six neighboring DNAs. Concurrently, however, DNA wrapping near the inner hole of DNA "donut" costs higher bending energies. The optimal toroidal radius *K* and thickness *M* obey the scaling relations $K \propto |E_0|^{-2/5} L^{1/5} l_p^{2/5}$ and $Th \propto |E_0|^{1/5} L^{2/5} l_p^{-1/5}$ [80], as functions of DNA-DNA attraction strength at optimal DNA density $E_0 = E(R_{opt})$ and DNA length *L*. According to Eqs. 2,3, in the presence of DNA-condensing ions DNA-DNA cohesive energy can reach E_0 =-(0.01÷0.1) k_BT per bp along the DNA-DNA contact. It plays the role of the surface tension controlling toroidal dimensions, see Fig. 5. The model reveals that DNA toroids become "fat" as the DNA persistence decreases and DNA-DNA attraction increases: torodial mean radius decreases and thickness grows.

Several theoretical models of DNA toroidal condensation with non-hexagonal and non-circular cross-sections have been proposed in the literature [83,84]. We also want to mention that, although locally the lattice of the wrapped DNA preserves the hexagonal symmetry to make best use of attractive inter-molecular contacts, the path taken by a *continuous* long

DNA upon wrapping into a toroid is still debatable [85,86]. Similar complications emerge for DNA packing inside viral capsids, see Sec. 11.

Perspectives. How stable are DNA toroids? Recent single-molecule optical tweezers manipulation experiments have enabled the researchers to decipher the physical mechanisms behind toroidal stability and gain some insights into DNA condensation dynamics [87]. In particular, a step-wise DNA unwrapping from toroids by applied forces was detected, corresponding to multiple DNA loops released from the condensate. The number of turns released is a function of applied tension of 1-10 pN and of salt-dependent DNA-DNA attraction. Theoretical statistical mechanics models of force-induced DNA unwrapping from DNA "donuts" have also been developed in recent years [88,89].

One intriguing perspective to enrich the morphology of DNA toroids observed in 3D is DNA condensation on positively charged 2D interfaces. Some analysis of deformations of model toroids on wetting/non-wetting surfaces was performed e.g. in Ref. [90], without accounting however for ES DNA-DNA and DNA-surface effects. Indeed, DNA condensation on 2D attractive surfaces (for instance, on CL membranes) is expected to follow different pathways and result in different final morphologies, as compared to DNA aggregation in 3D solutions.

Recently, a coil-globule DNA transition on *unsupported* CL membranes has indeed been reported in Ref. [91]. DNA globules of ~0.1÷0.4 μm in size emerge on membranes in 1:1 salt solely due to the presense of mobile positive lipids. They act as counterions for DNA, neutralizing its charge along the DNA-membrane contact. Although a precise morphology of condensates could not be resolved, the hydrodynamic radii of DNA globules were dramatically reduced with the increasing fraction of positive lipids in the membrane. Physically, some patches of positive lipids get bound to DNA deposited on the CL membrane, progressively wrapping around and compacting the DNA coil into a dense globule. The membrane deformations accompanying this process are vital, because *supported* membranes with the same lipid composition do not exhibit a coil-globule DNA transition [92]. Mixing and rearrangement of membrane lipids is also expected to play a role, similarly to the adjustment of lipid charges in DNA complexes with CL membranes, reviewed in the next Section.

4. DNA complexation with cationic lipid membranes

Structure of complexes. Self-assembly of CL membranes with oppositely charged biomacromolecules has been extensively studied experimentally for DNA [93], f-actin [94], microtubules [95], and some filamentous viruses [96]. Dense assemblies of DNA with CL-membranes are promising non-viral transfection vectors for gene therapy applications [97], successfully targeting nowadays several types of cancer [98].

Surface charge density on the CL membranes, +0.3÷1 e_0/nm^2, is often comparable to that of DNA and thus ES forces dominate their complexation into different phases. Depending on the fraction of cationic lipids, membrane flexibility, and lipid composition, dense well-ordered lamellar L_α^c [93] and inverted-hexagonal H_{II}^c [99] phases are commonly observed in experiment, Fig. 6. The H_{II}^c phases are preferred for artificially soft or intrinsically pre-curved membranes, when a tight cylindrical wrapping of membrane lipids around the DNA takes place. For the lamellar phase, ordered layers of parallel DNAs alternate with CL membranes (one DNA layer per one membrane), compensating the charge. Note that for f-actin, due to a mismatch in the charge densities, the unit cell of the lamellar stack consists of *two* negatively charged f-actin layers on *both* sides of a CL-membrane.

For DNA-CL complexes, most stable assemblies often occur at the *isoelectric point* of exact charge matching between DNAs and CL membranes [100]. The assembly process is accompanied by almost complete release of condensed counterions from the DNA and membrane surface [101]. Concomitantly, the translational entropy of these "evaporated" counterions is maximized. The DNA-DNA separations measured in DNA-CL complexes are in the range 25Å<R<60Å and they are often consistent with the picture of counterion-free assemblies.

The ES stabilization mechanism of DNA-CL complexes based on numerical solutions of the *non-linear* PB equation has been established a decade ago, in the model of non-fluctuating rod-like DNAs [102,103,104] and refined for more realistic setups in recent coarse-grained computer simulations [105,106,107]. For the lamellar complexes, a particular attention was paid to ES-driven DNA-mediated adjustment of CL charge density profile [108] and membrane undulations [109] that might help to improve DNA-membrane charge matching.

Model and results. Recently, we developed a similar ES model based on exact solutions of the *linear* PB theory, with the dielectric boundaries (DNA-solvent, membrane-water) and DNA helicity taken explicitly into account [110]. Within this approach, both for $L_\alpha^{\ c}$ phase with planar membranes and for $H_{II}^{\ c}$ phase with CL membranes wrapped around DNAs, the distribution of ES potential in electrolyte has been calculated. The variation of the complex ES energy was computed as a function of DNA lattice density and CL-fraction on the membrane. Both appear to exhibit a *non-monotonic* behavior, in good agreement with the numerical results of the non-linear model [103]. The energy minimum found from the model roughly corresponds to electro-neutral assemblies [110]. For the lamellar phase, the energy well near the minimum is attributable to ES compressibility of the DNA lattice. A scaling law for its modulus obtained, $B_{comp} \propto 1/R^2$, agrees well with the experimental data [111]. For $L_\alpha^{\ c}$ phase, the ES energy of DNA-induced undulatory membrane deformations can be included later on in more elaborate models.

The laws of DNA-DNA ES forces along and across CL membranes were also examined [110]. For instance, for two thin PE rods on a "salty" interface with mobile charges [112] one can get a power-law decay of ES interactions, in contrast to a nearly exponential decay of rod-rod ES screening known in 3D, Fig. 7. Confinement of electrolyte solution between the adjacent membrane layers also modifies the law of screening along the membranes. Namely, for a point charge, an exponential decay of ES potential at small distances d turns into $1/d^3$ power-law at large distances, when the electrolyte in inter-membrane spaces is rendered quasi-2D. The ES forces across a low-dielectric membrane are also renormalized non-trivially [110]. All these features affect the properties of DNA transversal and longitudinal correlations in DNA-CL lamellar phases, as measured in experiments [111].

The theory of duplex-duplex ES interactions enables us also to rationalize [110] the DNA-DNA separations measured in $L_\alpha^{\ c}$ phases in the presence of some di-valent cations (Mg^{2+}, Co^{2+}, Ca^{2+}). These are capable of triggering DNA condensation in 2D DNA-membrane system [113], but not in 3D solution. It was observed that at a critical concentration of divalent salt, ~20-60 mM, the DNA-DNA separations drop abruptly from ≈45Å for nearly electro-neutral complexes down to "universal" DNA compaction density with R≈29Å. Note again that DNA-DNA ES attraction is realized at these R, Fig. 2. Also, for multi-valent spermine^{4+} and spermidine^{3+}, the DNA-DNA distances in DNA-CL phase appear to be close to those measured in 3D lipid-free DNA condensates [114] and also agree with the predictions of our theory of DNA-DNA ES interactions, Fig. 8. One can conclude that 2D geometry of DNA-CL $L_\alpha^{\ c}$ phase facilitates inter-DNA counterion-mediated attraction, rendering ES attraction possible for divalent cations and at DNA charge neutralization fractions measured to be $\theta \approx 0.63$ [113], lower than the Manning estimate of $\theta_M = 0.8\text{-}0.85$.

Applications. A perspective for future research is to enrich the physical understanding of DNA release from sub-μm sized DNA-CL complexes and of their translocation into the cell cytoplasm across negatively charged cell membranes [98,115]. Both processes are necessary for efficient gene delivery, with the transfection efficiency of DNA-CL complexes still remaining low compared to viral-based gene carriers [98]. It is known e.g. that spermine and spermidine not only compactify DNAs in DNA-CL complexes, but can also trigger a DNA release from them via DNA condensation into dense globular/toroidal aggregates in solution. Another pathway for DNA release is designing lipid membrane being unstable in particular cellular cytoplasmatic environments (addition of special "helper" lipids).

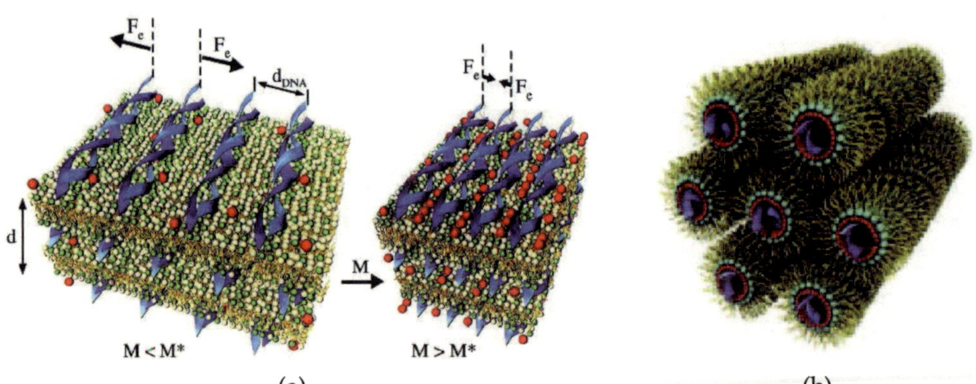

Fig. 6. Schematics of 2D DNA condensation in the lamellar DNA-CL-membrane phase with divalent cations (a). Inverted hexagonal $H_{II}^{\ c}$ phase (b). Part a) is reprinted from Ref. [113], copyright 2000-NAS, USA.

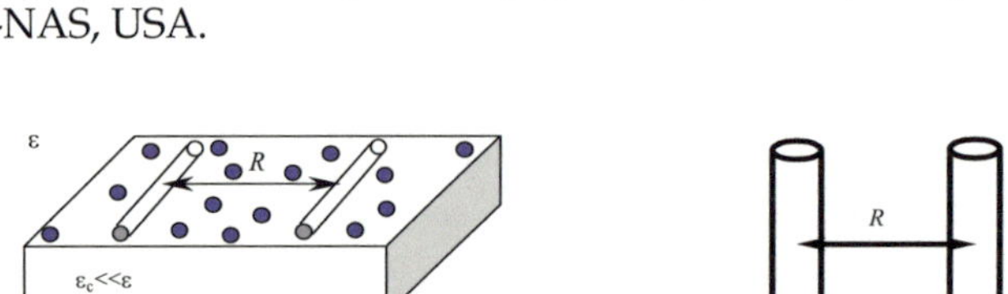

$$E_{2D}(R) = k_B T \frac{l_B}{b^2 \kappa_s^{\ 2} R^2} \qquad E_{3D} = 2 k_B T \frac{l_B}{b^2} K_0(\kappa R)$$

Fig. 7. Energy density of rod-rod ES repulsions along a "salty" lipid membrane with the inverse Debye length $1/\kappa_s$ and in 3D electrolyte solution.

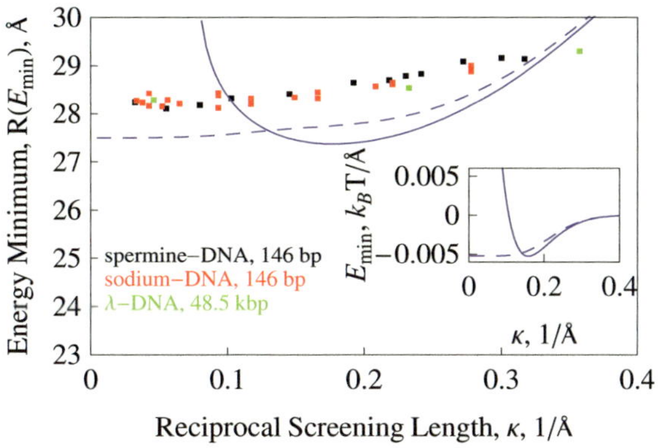

Fig. 8. Optimal DNA-DNA separations measured in dense DNA precipitates with spermine^{4+} experimentally [114] (dots) and predicted theoretically [110]. The theory curves with the Donnan saturation (dashed) are more realistic, while the solid curves are for external salt levels also inside the DNA lattice. Multivalent cations only affect the screening in the model: DNA charge fraction (1-θ) stays constant. The inset depicts the ES DNA-DNA energy in the local energy minimum. Parameters: θ=80%, f=0.3.

5. DNA cholesteric phases

Properties of DNA twisted phases. DNA chirality on the nano-scale manifests itself via DNA-DNA interactions in a formation of twisted, ~160-350 mg/ml dense, LC DNA phases on the micro-scale [116,117]. Cholesteric phases also emerge upon assembly of other bio-helices, e.g., collagen fibers, filamentous viruses [118] and guanosine [119]. Some poly-peptides feature LC phases too. Nature uses the ability of DNA to form chiral phases for packing the genomes in some bacteriophages [120], bacteria [121], and in sperm of many vertebrates [122].

Typically, *left-handed* DNA LC phases are detected, with the cholesteric pitch of P~1÷4μm for ~150bp long nucleosomal DNA fragments [123]. The pitch dependences on the [salt], temperature T, DNA lattice density, and external osmotic stress are however non-trivial functions. For example, the pitch P decreases at higher [NaCl] in the range 0.2-1 M [123]. It reaches P~20μm for DNA phases with some multivalent cations added [124] and can be reversed by addition of short basic polymers such as poly-Lysine and chitosan.

A number of theories of DNA cholesteric ordering have been developed based on the helical nature of DNA charges [125] in order to rationalize these and many other observations. Some geometrical models imply that right-handed cholesteric pitch is favored by a steric hindrance of DNAs [126], while left-handed phases should originate from ES interactions [127]. Other, purely ES models, treat the DNA charge helicity explicitly and predict a *right-handed* twist direction to be favored for two right-handed DNA duplexes in a close contact [59]. Such twisting direction ensures a more parallel and more ES-beneficial arrangement of negative phosphate strands of one partially neutralized DNA with the "strands" of adsorbed counterions on the neighboring helix. This fact results however in a *right-handed* twist of DNA cholesterics, opposite to a number of experimental observations.

Model and its conclusions. Based on the theory of ES interactions of two skewed DNAs [58], we examined the ES stability of DNA cholesteric phases calculating the strength of DNA-

DNA azimuthal correlations [128], Fig. 9. The DNA "triad model" was implemented, in the ground-state, with no fluctuations, and with a perfect DNA azimuthal register on the lattice [59]. The theory predicts a *non-monotonic* pitch dependence on the DNA density for ≈150bp DNA fragments, in agreement with experiments [123]. Also, the range of DNA densities of R=35-45Å predicted is often close to the measured stability domains of DNA cholesterics, Fig. 10. This indicates that long-range ES forces, rather than short-range steric hindrance of the grooved DNA surfaces, are likely to be responsible for DNA LC ordering.

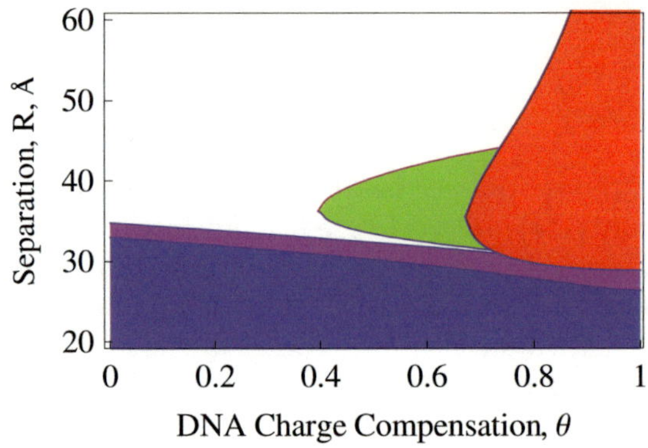

Fig. 9. Stability domains of DNA cholesteric phases. Azimuthal correlations of 150 bp DNAs are strong inside the green domain, as predicted by the theory [59] with the Donnan equilibrium [129]. The energy of azimuthal DNA rotation on the hexagonal lattice exceeds $k_B T$ in the green region. The LC twist elastic constant is $K_{22} < 0$ inside the red domain, the DNA azimuthal rigidity constant is $k_\phi < 0$ for magenta and blue domains (these regions are non-physical). Parameters: $\lambda_D = 7$ Å, f=0.3.

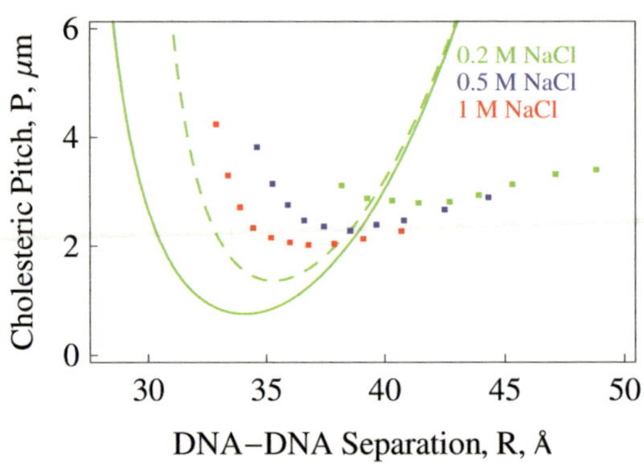

Fig. 10. The value of DNA cholesteric pitch, as calculated from the theory [128] with (dashed) and without (solid curve) the Donnan equilibrium in the DNA lattice. DNA assembly is pressurized to retain a proper [DNA]. Experimental points for DNA LC phases of 146 bp DNAs are taken from Fig. 5a of Ref. [123]. Parameters: f=0.3, θ=0.65, and $\lambda_D = 7$ Å that corresponds to ~0.2 M of NaCl.

Strong DNA-DNA azimuthal correlations are vital for formation of DNA cholesterics. Our ES model [128] predicts that these correlations vanish both at small and large DNA lattice densities. In the first case, it is due to the decay of DNA-DNA ES interactions, Eq. 3, while in the small-R region the inherent azimuthal *frustrations* of DNA-DNA potential (Eq. 4) destroy DNA orientational order. The existing ES theory of DNA cholesterics [59] has been modified to incorporate the Donnan electro-chemical equilibrium of ions [129] in DNA lattices. This effect appears to be particularly important at low [electrolyte], as follows from the equation for renormalized screening length inside a DNA phase [128]

$$\lambda_D^{Don} \approx \lambda_D \left[1 + \left(\frac{4 l_B \lambda_D^2 (1-\theta)}{b\left(R^2\sqrt{3}/(2\pi) - a^2\right)} \right)^2 \right]^{-1/4} . \tag{5}$$

Thus, in dense nearly electro-neutral DNA assemblies the DNA lattice density, rather than the bulk [salt], dictates the ionic conditions of solvent between the DNAs, Fig. 11.

Despite a good agreement for the pitch value, the direction of winding of DNA cholesteric layers and the shift of stability domains at different [salt], see Fig. 10, cannot be rationalized by this ES theory in its current form. Right-handed pitch is anticipated at relevant DNA densities of R=35-45Å, with a possible change to left-handed rotation for very dense DNA packing or atypical counterion patterns on DNA [125]. In the model of dense LC phases with thermally undulating, rather than straight DNAs, a right-to-left pitch inversion might originate from an enhanced contribution of ES image forces. The latter can favor the opposite sense of DNA-DNA crossing, compared to the direct duplex-duplex ES forces. This conjecture requires a detailed future analysis.

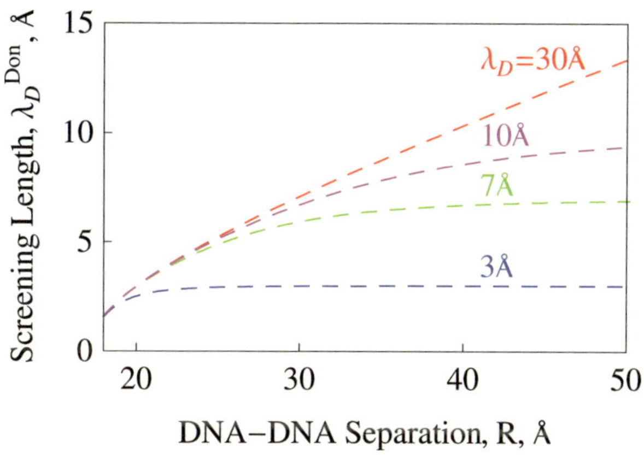

Fig. 11. Effective screening length in dense DNA assemblies for different [salt] in the bulk, as calculated in the cylindrical cell model according to Eq. 5.

Right- and left-handed phases. A major break-though in understanding of LC DNA phases was accomplished recently in Ref. [130]. Namely, an *inversion* of cholesteric handedness in dense LC DNA phases build from ~6÷20 bp short DNA fragments has been revealed. Subtle changes in DNA sequence and fragment length were shown to trigger this inversion. These very short DNA fragments stack on each other to form elongated DNAs with a sequence-specific 3D structure [131,132]. For a stacking procedure that generates more regular helical DNA strands, predominantly left-handed LC phases were observed, similarly to those for

≈150 bp DNA fragments as well as for kbp-long DNAs. Right-handed phases were observed for short sequences and more azimuthally flexible stacking connections of DNA nano-rods.

The analysis of data (for a variety of stacked DNA oligomers, at different [DNA] controlled by sample dehydration, etc.) enabled the authors to reconcile the results in terms of a single physical parameter [130]. Namely, for DNA lattices with isotropic-nematic phase transition below ≈620 mg/ml, the left-handed DNA cholesterics are formed. DNA sequences that experience this transition at larger [DNA], give rise to right-handed LC phases. Thus, for DNA-DNA distances of $R \leq 32$Å the right-handed DNA-DNA crossings seem to be favored [133], whereas LC twist is right-handed for DNA fragments shorter than 14 bp [130]. Some of these experimental trends have been supported in recent simulation study [134].

Also, a general trend was detected [130] that shorter DNA oligomers form cholesteric phases with a shorter, sub-μm pitch. This fact is consistent with the ES DNA theory [59] being also akin to $P(L)$ variation for longer DNAs [135]. For shortest DNAs, the pitch of ~0.3 μm was detected, much smaller than 2-4 μm for ~150 bp fragments. Systematic analysis of [salt] and T-dependence revealed also several unexplained features. Some sequences exhibit e.g. a pitch reversal at [DNA] of 620 mg/ml. Also, for the majority of DNA oligomers, the pitch increases with T indicating an unwinding of cholesteric structures, regardless of their handedness. For many sequences, the pitch was almost insensitive to [salt], contrary to a strong P([salt]) dependence measured for ≈150 bp DNAs [123]. All in all, this detailed investigation of twisted phases of stacked nano-DNAs [130] has enriched enormously a widely accepted view of "left-handed-only" DNA cholesterics, challenging future theoretical investigations of DNA twisted phases.

6. DNA-mediated ES interactions of nucleosomes

DNA compaction. NCP-mediated DNA compaction in eukaryotic chromosomes ensures enormous compactification "power" often required to pack meter-long genomes in ~μm-sized cell nuclei [19]. This process takes place on several hierarchical levels, with the initial step being the 30-nm chromatin fiber, an organized array of NCPs connected by a continuous DNA. Structure of this fiber is sensitive to many biological factors such as ionic environment, DNA linker length, concentration of H1 linker histones [136], charge state of histone tails, etc. It is still debated what conditions favor solenoidal [137] vs. zig-zag [138] arrangements of NCPs in chromatin fibers. *In vitro*, recent fiber reconstitution experiments provided a great deal of information about inherent stability, NCP linear density, and diameter of chromatin fibers for various lengths of DNA linker [139]. The energetics of DNA-histone [140], H1-NCP, and NCP-NCP [141] contacts is vital for physical understanding of chromatin fiber structure, stability and functioning.

Similarly to unwrapping of DNA toroids that enabled to rationalize their stability [87], recent measurements of force-induced stretching of chromatin fibers revealed the NCP-NCP cohesive energies of ≈3.4 [142] and later of ≈14 k_BT [143], depending on concentrations of mono- and di-valent cations. Ionic conditions dramatically influence also the forces required to disrupt the NCP-NCP contacts in chromatin fibers (typically ~2-5 pN) and to induce a DNA unwrapping from NCPs (~6-20 pN). These cohesive energies guarantee the fiber stability, but also allow for some "unwrapping plasticity" and "breathing dynamics" of complexed NCPs, both required for DNA transcription to take place.

Top-to-bottom NCP-NCP stacking contacts are likely to be hydrophobic, governing formation of NCP columns in NCP crystals, semi-crystalline NCP phases, NCP multi-layered helices, etc. The side-to-side contacts, with the wrapped DNAs being often in close contact, on the contrary, are likely ES in nature, Fig. 12. Close similarities in condensation and re-solubilization of DNAs and NCPs support this statement. These [salt]-dependent ES contacts control the formation of mesophases of isolated NCPs *in vitro* [144].

The stability of dense NCP phases is controlled by side-to-side DNA-mediated NCP-NCP contacts. In these phases, the DNAs on contacting NCPs are separated by only 5-15Å of electrolyte, that is typically within one Debye length λ_D. For the NCP bilayer phase [145], the NCPs are oriented in columns so that their dyad axes point on average perpendicular to the bilayer plane and the NCP sides with 2 DNA turns are buried inside the bilayer. The NCP azimuthal orientations are frustrated and distributed within ±≈35° from this preferred direction [145]. Stronger DNA-mediated contacts of NCPs *inside* the bilayer govern its formation, being likely responsible for peculiar NCP azimuthal frustrations observed.

Fig. 12. Schematics of inter-nucleosomal interactions, with the positive histone cores shown in blue and super-helically wrapped DNA depicted in red. This in-plane orientation of NCPs corresponds to a 2:2 DNA-DNA contact.

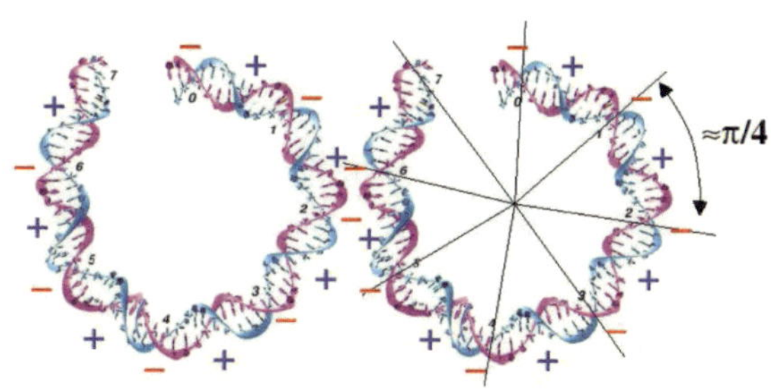

Fig. 13. Schematics of NCP-NCP interactions modulated by charge periodicity of wrapped DNA. Zipper-like charge motif is viewed along the DNA superhelical axis, for a referential counterion adsorption into the DNA major groove. Data of NCP 1aoi.pdb structure was used. The image is reproduced from [147] with permission of IOP.

Model of inter-NCP ES forces. Crystal structures of NCPs show ≈8 full DNA helical turns per one superhelical turn of the wrapped DNA. DNA turns in NCPs are separated by the same "magic" $R≈25$Å. At these distances parallel DNAs attract each other in dense assemblies with $MnCl_2$, Fig. 2, and they are close to equilibrium separations in DNA toroids, Sec. 3. An important ingredient of the model is the structural fact that two turns of DNA wrapped in NCPs bear a strong positional *register* forming a "super-groove"[146]. As we have seen in Sec. 2, the DNA in electrolyte solution exhibits a distinct pattern of alternating charges along its axis (negative phosphates vs. positive condensed cations). Thus, this positive/negative DNA *charge zipper* along the side-to-side contact of two in-plane NCPs modulates their azimuthal interactions with the period of $\approx \pi / 4$, see Fig. 13.

For in-plane NCPs with parallel axes, we demonstrated that this azimuthal modulation gives rise to quantization of NCP orientations in nucleosomal bilayers with periodicity of ≈45° [147]. Azimuthal optimization of side-to-side contacts of NCPs in bilayers resembles the azimuthal adaptation of short DNA fragments in columns of nano-DNAs in the cholesteric phases, Sec. 5. Both effects likely originate from DNA helix-helix ES interactions.

Generally speaking, the DNA contribution to NCP-NCP ES energy is 4 times stronger for the NCP sides with 2 DNA turns as compared to the NCP sides with only 1 DNA turn. For the in-plane NCPs, we calculated the ES forces and attraction-repulsion phase diagram, see Fig. 14. We implemented a simple model of ES double-layer repulsion for the histone cores, modeled as uniformly charged spheres with charge Q. For the DNA part, the Derjaguin approximation was used to compute the ES forces between the bent DNA duplexes, interacting locally according to Eqs. 2,3.

For a typical histone charge Q=+220e_0, at physiological salt conditions, the model predicts for the 2:2 DNA-DNA contacts the maximal NCP-NCP attractive forces of 2 pN for $\theta = 0.8$, f=0.3 (typical parameters used in the theory, as in Fig. 14). The NCP-NCP attraction reaches 8 pN at $\theta = 0.9$, f=0.3 (better DNA charge neutralization), and even 60 pN at $\theta = 0.8$, f=0 (stronger DNA-DNA attraction due to binding of cations into the major groove [50]). As the histone positive charge grows, the DH repulsion of NCP cores overwhelms the DNA-DNA ES attraction. Thus, at larger Q/e_0 values, the NCP-NCP attraction region at 25-35Å between DNA fragments along the NCP-NCP contact disappears, see the inset in Fig. 14.

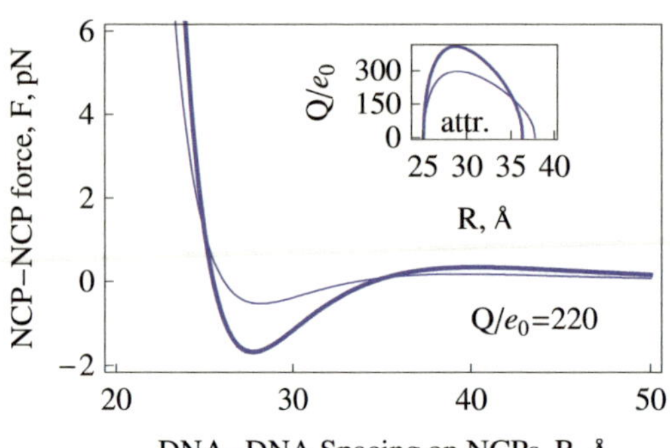

Fig. 14. ES force between two NCPs with parallel axes. Optimal NCP azimuthal alignment is assumed: DNA-DNA attraction. Here R is the distance between DNA axes on NCPs contacting side-to-side (R=2a=18Å is direct DNA-DNA contact). Thick and thin curves are for 2:2 and 1:1 DNA-DNA contacts, respectively. ES NCP-NCP attraction-repulsion diagram is shown in the inset. Parameters: $\theta = 0.8$, $f = 0.3$, $1/\kappa = 7$ Å.

DNA-mediated ES attractions of 2:2 vs. 1:1 DNA sides of NCPs are likely to trigger the bilayer formation and NCP azimuthal frustrations. Another possibility is that histone tails bridge neighboring NCPs in bilayers in azimuthally dependent manner [145]. At typical conditions, the range of DNA-mediated and tails-mediated NCP-NCP interactions might indeed overlap, being thus hard to distinguish. And, the histone tails also follow the symmetry of DNA in NCPs, protruding into solution through the aligned minor grooves of the DNA superhelix.

To reconcile numerous observations for semi-dense NCP phases, NCP crystals, and chromatin fibers, a rigorous theory of ES NCP-NCP interactions is to be developed in the future. For arbitrary orientations of NCPs in space, one has to take into account the helicity of DNA charges, a heterogeneous distribution of histone charges on the side and top/bottom NCP surfaces, a low-dielectric core of histones and the DNA, DH ES screening by electrolyte, and the counterion separation-dependent condensation. All these effects make such a theory a formidable problem of the mathematical physics, even within the linear PB theory.

7. Homology recognition of DNA sequences

DNA structure. So far, ES forces between ideally-helical parallel or skewed DNAs were described. The locality of intermolecular potential gives rise to the same ES energies for DNA fragments hom in sequence, but not ideally helical. Below, the sequence effects on DNA-DNA ES forces are overviewed. To save space, we dwell here on several subjects only, addressing the reader to an excellent recent perspective [67] that covers all aspects of ES DNA-DNA recognition and also suggests biological phenomena where it is of potential importance. One immediate application of the theory is to provide a physical rationale [148] for recognition and pairing of hom genes on genomic ds-DNAs [149,150] during cell division.

The physical mechanism of ES DNA-DNA sequence recognition was pioneered in Ref. [53] for parallel torsionally rigid DNAs and later extended for duplexes with a realistic value of torsional rigidity [55]. The ES recognition emerges in the model solely due to the inherent bp-specific non-idealities of DNA helix, as extracted from the analysis of structural data on DNA-DNA and DNA-protein crystals [151]. In particular, the DNA bp twist angles are known to exhibit a strong variation [52,152] fluctuating in a range of 28-40° that gives rise to ≈36±5° angle deviations.

In the theory, these variations for a randomly-sequenced DNA form the *helical coherence length* that is $\lambda_c = H/\left(10\,\Delta\Omega^2\right) \approx 45\,\text{nm}$ at a typical value $\Delta\Omega \approx 5\,^{\circ}$. This length controls the degree of DNA non-idealities that strongly affect DNA-DNA ES forces. A finite DNA twist persistence length $l_{tw} \approx 75\,\text{nm}$ allows for some interaction-induced DNA torsional adjustments to take place. These restore to some extent the DNA helical register along such sequence-unrelated DNA fragments [55].

Results. We define below the recognition energy $\Delta E(L)$ as the difference in ES interaction energy for hom and randomly sequenced DNA fragments of length *L*. For torsionally *rigid* DNA fragments, with azimuthally free ends, in the leading a_1-approximation we get [53,70] $\Delta E(L) \approx a_1 \lambda_c \left[L/\lambda_c + 2e^{-L/(2\lambda_c)} - 2 \right]$. In the opposite limit of torsionally adaptable, *very soft* DNA duplexes, the recognition is also described by a simple formula [70] $\Delta E(L) \approx a_1 \dfrac{\lambda_t}{2\lambda_c} \left[L - \dfrac{\lambda_t}{2}\left(1 - e^{-2L/\lambda_t}\right) \right]$, where $\lambda_t(R) = \sqrt{C/\left[2a_1(R)\right]}$ is the *R*-dependent DNA torsional adaptation length. For a standard value of DNA twist modulus, $C = 750 k_B T$ Å, the exact analytical expressions for $\Delta E(L)$ are however quite cumbersome [55]. Roughly, at relevant parameters, the theory predicts that two DNA fragments with unrelated sequences attract each other nearly half as strong as two hom DNA sequences do, see Fig. 15a.

The ES recognition energy predicted grows linearly with the length of DNAs in contact [55], resembling thereby some properties of DNA hom recombination *in vitro* in the absence of specific DNA-pairing proteins, Fig. 15c. The recognition energy exceeds several $k_B T$ for closely aligned DNA fragments of ~200-500 bp in length. This energy is large enough to ensure a stable pairing of hom DNA segments at ambient temperature. It can also be sufficient to trigger unpairing of DNA single strands, required as initial step of hom recombination.

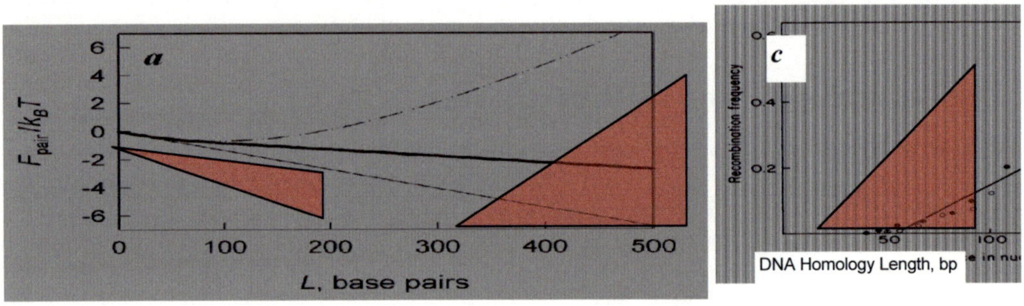

Fig. 15. a) Computed ES interaction energy in a pair of hom DNAs (thin), randomly-sequenced torsionally rigid DNA fragments (dot-dashed), and randomly-sequenced DNA fragments with a realistic twist rigidity C (solid curve). b) The corresponding DNA-DNA ES recognition energy. Parameters: R=30Å, $\lambda_D = 7$ Å, θ=0.8, f=0.3, $l_{tw} \approx 750$ Å. c) Measured frequency of hom recombination events in T4 phage [153]. It shows a minimal length of DNA homology of ≈50 bp necessary for recombination to start and a linear growth of frequency with DNA homology length. This resembles a linear growth of the recognition energy for long DNA sequences in b). The images are reprinted from Ref. [55] with permission of the ACS and from Ref. [153] with permission of Elsevier.

For pulling two ds DNAs one over another, with the hom bp domains in them, the energy well for ES recognition has recently been theoretically computed [148]. For very closely juxtaposed DNAs at R=30Å, the recognition energies of up to 5-10 $k_B T$ and pinning forces near the bottom of the well of ~2 pN were predicted for typical DNA parameters, Fig. 16.

Sequence-specific DNA recognition and pairing for intact duplexes was indeed observed for yeast hom DNA chromosomal loci, in the absence of any recA-family proteins [154]. It was attributed to some sequence-specific DNA-DNA forces, capable of initiating and maintaining a proximity of hom DNA fragments. Recently, several experimental techniques have been utilized to elucidate the properties of DNA-DNA hom pairing in denser arrangements [155,156,157]. In one study on dense DNA cholesteric spherullites, the segregation of ~300 bp DNA fragments with identical bp sequences into separate LC populations has been clearly identified [156] (telepathic DNAs). This offered a first proof of direct DNA-DNA recognition, based solely on DNA bp sequence information. The effect was attributed to more favorable ES interactions of hom DNA fragments as compared to non-related ones. ES DNA-DNA bp-specific forces [148], are likely to be responsible for the observed segregation of hom sequences at these relatively high [DNA] corresponding to R=32÷40Å.

In another study, single-molecule magnetic tweezers measurements revealed an efficient sequence-specific pairing of λ-ds-DNAs with hom regions longer than ~ 5 kbp, at [salt] and [DNA] close to those *in vivo* [157]. The paired structures of hom DNAs were sheared by $F \approx 10$-20 pN forces and pairing was more profound in the presence of $MgCl_2$, indicative of ES nature of this effect. Some other properties, such as a strong enhancement of pairing efficiency with [simple salt] up to 1 M as well as a non-monotonic T-dependence favored however rather some non-Coulomb origin of pairing forces. Also, in experiments, the precision of hom DNA-DNA register along the DNA pair measured is often ~2-5 μm, much larger than the width of the recognition energy well in the model, $\sim \lambda_c \approx 10 \div 50\,\mathrm{nm}$ [148], see Fig. 16. This width in experiments is also independent on the length of the paired hom DNA segments. Future developments of this highly promising technique might provide more information about the axial proximity of paired hom DNA fragments and thus enable us to estimate the effective "range of action" of these sequence-specific DNA-DNA forces.

To summarize, we do believe that helix-specific ES forces can govern DNA-DNA sequence recognition in dense DNA phases [156], at high [DNA] and suppressed DNA fluctuations. DNA-DNA homology associations *in vivo* are however often maintained at much larger separations and take place between fluctuating DNAs [158]. The pairing remains efficient at DNA-DNA distances of 100-300 nm, much longer than the Debye screening length that limits the action radius of ES forces. DNA-DNA hom pairing should thus also involve a recognition mechanism other than the direct ES forces [159], probably recruiting proteins for protein-mediated homology-specific DNA-DNA contacts.

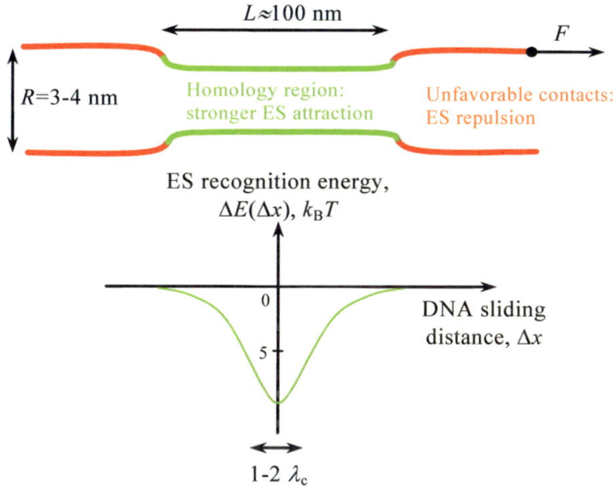

Fig. 16. Pictorial shape of the recognition energy well for sliding of two DNAs with a hom domain, as obtained from Eq. (1) of Ref. [148] for rigid DNAs. DNA hom segments are marked in green; non-hom sections are in red. Hom fragments are pinned near the well bottom by a stronger ES attraction, relative to the rest of DNA. Parameters: $\lambda_c = 100$ Å, R=30Å, $\theta = 0.8$, $f = 0.3$, $1/\kappa = 7$ Å that give $a_1 \approx 0.015\ k_B T/$Å. The helical coherence length for DNAs in solutions and wet fibers λ_c = 10-20 nm, is much shorter than in DNA crystals, λ_c = 50-100 nm [57], where the helices are "straighten" by mutual interactions.

8. Close-range DNA-DNA ES friction

Modern nano-tribology applications necessitate a detailed understanding of frictional forces between bio-molecules on the nano-scale [160,161]. For DNA, recent advances in single-molecule manipulation techniques has allowed measuring the forces required to pull one DNA over another one in a tight superhelical DNA ply, the dual optical trap. Tight winding of two DNAs in the ply can facilitate their interactions [162]. Upon shearing the ply, in the presence of DNA-associated DNA-bridging H-NS proteins, the frictional forces up to ~25 pN were detected [163]. They emerge from disruption of DNA-protein-DNA bridges formed every several *H* along the ply. Also, when some small proteins bind to the ds-DNA and sterically impede DNA pulling, the friction of ~2-5 pN was detected. For the "bare" DNAs, one could expect that inherent DNA helicity on *H*=3.4 nm scale might itself generate some friction. With the same apparatus, no measurable friction was detected however [164]: surprizingly, the forces remained <1 pN, independently on the length of DNA ply, DNA pulling speed, and the presence of DNA-condensing (spermine^{4+}) cations. Diameters of DNA plectonemes in experiments were estimated to be ~5-10 nm.

In this and the next Section, we discuss two manifestations of sequence-dependent DNA-DNA interactions considered in Sec. 7, for DNA-DNA friction and DNA melting. Using the theory of DNA-DNA ES forces, Sec. 2, we examined different regimes for DNA-DNA nano-friction depending on the character of DNA sequence [165]. For ideally helical non-fluctuating and closely juxtaposed DNAs, the ES friction emerges due to spatial correlations of ES potential along DNA surfaces, Fig. 17. At relevant [salt], these correlations are only pronounced in the first ~10Å from the DNA surface. ES frictional force in this regime oscillates with the period of *H*=3.4 nm, while its magnitude grows linearly with the length of DNA *L*. Namely, the force of static friction is $F_{fr} = 2\pi a_1 L / H$. For slow DNA pulling, this gives rise to a *stick-slip motion* on the nano-scale [165]. The friction however remains rather low. Even for very tight DNA plies, with thickness of ~40 Å and parameters favoring DNA-DNA attraction (large a_1), the upper estimate for frictional force in a ply of $N = L / H = 10$ DNA turns long is as small as ≈4 pN.

Several effects are likely to reduce this upper limit. It is the case for pulling non-ideally helical DNAs (random bp sequence). For such DNAs, the "corrugations" in DNA helical structure progressively accumulate with the length [53] and ES potential variations along the DNA-DNA contact become de-correlated, as discussed in Sec. 7. This, in turn, strongly impedes ES friction that attains in this limit an exponential decay with the pulling distance of one DNA with respect to another one. As ES forces decay exponentially with *R*, it is not surprising that for DNA plies that are typically much thicker than 40Å, being formed by fluctuating DNAs with quasi-random sequences, no measurable friction has been detected in DNA-pulling experiments [164].

The situation might however change, when tight DNA plies are realized (by larger static stretching forces applied to DNA ends) and for DNA sequences with some degree of bp homology. One important example is DNAs fragments artificially designed to contain *repetitive bp hom blocks* with the length of ~50-300 bp. Then, one could expect some homology-mediated DNA pinning events upon mutual pulling of DNAs at the positions when these hom blocks on two DNAs overlap. The theory predicts [148] that these pinned states have a measurable half-width of $\sim \lambda_c$, see Fig. 16. Therefore, every bp-block pins with its hom partner on another DNA and their action along the chain accumulates enhancing the magnitude of DNA-DNA friction (i.e., the force to remove the system from the favorable state of fully overlapping hom blocks increases). This renders the detection of such pinning mode amenable for the current experimental technique [164] with the resolution of several repeats H. We would like to encourage the exprimental groups to check whether such blocky-homologous DNAs experience different frictional forces in tight DNA plies. On the contrary, the resolution of at least $H/2$ is necessary to probe the predicted above ES DNA-DNA friction F_{fr} on the scale of DNA helicity of 3.4 nm.

One potential aplication of DNA-DNA friction discussed is on DNA ejection from ds-DNA phages, when densely packed DNA strands have to slide passing each other upon reorganization of DNA layers inside the capsid during DNA compactification and ejection.

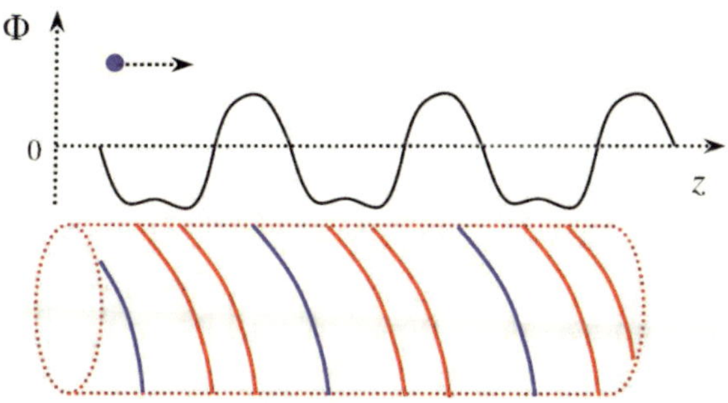

Fig. 17. Schematics of ES potential barriers $\Phi(z)$ near the B-DNA surface [166]. Negative DNA phosphate strands are shown in red, the counterions adsorbed in the DNA major groove are depicted as blue helices.

9. DNA melting in columnar assemblies

Upon heating up to ≈50-100°C, depending on GC-content and bp sequence, DNAs melt cooperatively in solution and their strands separate. Being thoroughly studied at low [DNA] [167,168,169], DNA melting in dense assemblies, when intermolecular forces become comparable to the internal DNA binding energies, remains not completely understood. The effects of DNA sequence on DNA melting in hexagonal assemblies are discussed below, within a simple thermo-dynamical model [170]. Namely, we predict that melting of hom DNAs is inhibited, while ds-to-ss DNA transition for un-related DNA sequences is facilitated.

In particular, it is straightforward to show that under the conditions favoring duplex-duplex ES attraction, ideally helical hom DNAs melt at higher T due to a stabilization of DNA helical regions by mutual interactions. For hom DNA fragments, the model predicts a rise of the melting temperature T_m (typically by 3-10° at R=23-28Å between DNAs in the assembly) and more cooperative DNA melting transitions. The shift of T_m scales with the strength of attraction, namely $\Delta T_m \propto 3|a_0 - a_1|$. It can thus be controlled in future melting experiments in dense DNA phases via addition of attraction-mediating cations, e.g., counterions with enhanced binding into the DNA major groove and propensity to induce DNA aggregation.

Also, the model predicts a change in the character of the melting transition, from the second to the first order at a critical strength of DNA-DNA ES attraction [170], Fig. 18. Then, at the melting point the fraction of ds helical DNA regions exhibits a discontinuous change. At the isotropic- cholesteric DNA transition, at moderate [DNA]≈100-150 mg/ml, a clear indication of such T_m jumps by several degrees was observed and quantified already long ago [171].

Recently, for dense aggregates of 10-50 nm GNPs linked by short DNA fragments [172] an extremely sharp DNA melting transition has been monitored, with a width of transition of 1-2°C, much smaller than ~10-20°C for melting of the same DNA in solution. This *enhanced cooperativity* of melting was attributed to short-range DNA-DNA interactions that trigger an accumulation of ions from electrolyte in the overlapping double layers around DNAs [173], in a Donnan-like fashion. For dense DNA bundles connecting DNA-functionalized GNPs [174], with DNA-DNA distances $R = 25 - 40$ Å, a higher local [salt] near DNAs are realized, which in turn tends to stabilize the ds-DNA state and increase T_m . The effective growth of T_m at these [DNA] was calculated to be 5-20° [173], based on a linear increase of T_m with log[[salt]] [172]. Once the melting of DNA bundle connecting GNPs starts, it progressively releases the excess counterions thus destabilizing the remaining ds DNA links. This might cause a sharp melting of the entire GNPs-DNAs assembly, as detected in experiments [172].

Our DNA melting model for dense aggregates of identical DNA helices [170] does account for the Donnan equilibrium in the DNA lattice, that would give a corresponding [salt]-induced rise of T_m [175]. The additional T_m shifts illustrated in Fig. 18 are however solely due to ES attraction of hom ds DNAs that appears also to be capable of inducing abrupt changes in the average DNA helicity.

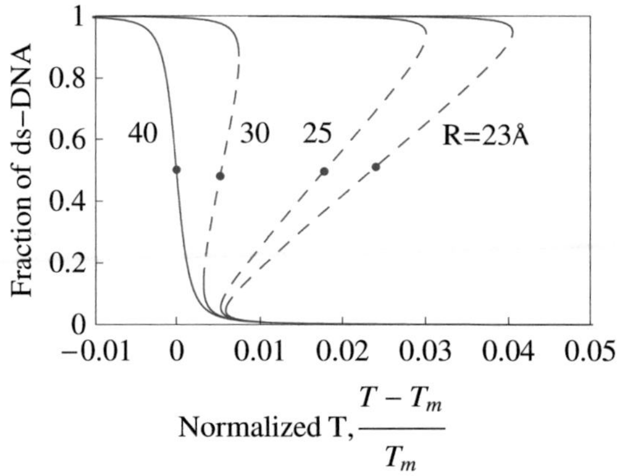

Fig. 18. Melting profiles predicted for long hom DNAs in the hexagonal assembly at varying DNA density. Dots on the curves indicate the T_m at which an abrupt charge in DNA helicity occurs (between the stable branches of the melting curve, due to a Z-shaped DNA melting isotherm realized at large enough DNA-DNA attraction). Parameters are the same as in Fig. 16. The figure is reprinted from [170], subject to ACS-2005 Copyright.

For ds DNA random in sequence, pressurized externally to form dense assemblies, one can expect on the contrary a destabilization of the helical state [170] by DNA-DNA ES length-dependent interactions overviewed in Sec. 7. The physical mechanism here is more formidable. Namely, the melted ss DNA domains/bubbles are going to be created on ds DNAs to optimize *unfavorable ES repulsions* predicted for bp-random DNA stretches longer than 1-2 λ_c [55], see the dot-dashed curve in Fig. 15a. The melting transition becomes less cooperative in this case. An experimental support of some of these findings might come from T_m -measurements in strongly confined wet DNA films [176].

Lastly, recently, I contributed to the invention of a method and apparatus to detect DNA melting and hybridization events with the help of bio-nano-sensors functionalized with dense DNA lattices. The situations of direct DNA attachment to the sensor surface and DNA deposition via GNPs have been considered. For the former, the sensor signals recorded upon DNA hybridization were rationalized [177] based on the theoretical model of redistribution of mobile counterions in spaces between DNAs (Donnan equilibrium). For GNP-DNA deposition, a more detailed ES screening model was developed [178]. This second detection setup allowed us to systematically monitor the change in the sensor response depending on mismatches in bp composition between probe and target ss DNA sequences. Such sensitivity is vital for a number of biological and biomedical applications involving detection of DNA sequences. This technique might also enable a detection of sequence-specific effects in DNA melting predicted above.

10. DNA-protein ES recognition: Models and reality

Protein-DNA recognition. Despite enormous experimental and theoretical efforts, the recognition laws of DNA-binding proteins and their cognate sites on ds DNA remain quite obscure. Geometric shape complementarity of proteins with DNA grooves and protein-DNA charge matching often drive the complex formation. Protein structures and their DNA recognition domains are extremely diverse that makes it hard to establish some universal rules of DNA-protein recognition. Several types of interactions can contribute to protein-DNA binding, with the ES and HB contacts being often the dominant ones.

The ES forces are known to dominate a non-specific binding mode for a number of DNA-protein complexes, e.g., weakly bound lac repressor [179]. DNA-binding domains of many relatively small proteins contain positively charged patches that ensure their ES attraction to the DNA. For large protein assemblies, the situation is often quite similar. For RNA Poly II, for instance, a strongly positively charged cleft is identified in the crystal structure along the path taken by DNA-RNA hybrid upon transcription [24]. For ribosomes, the basic residues are also located in protrusions of the structure expected to be involved in binding of tRNA/mRNA during translation [23].

Indeed, Lys^+ and Arg^+ residues in DNA-protein complexes are often located only several Å away from DNA phosphates, Fig. 19. ES DNA-protein contacts are however often believed to bear little specificity to DNA sequence [180], merely providing a proximity of proteins to the DNA and allowing more sequence-specific and orientation-dependent HB contacts to recognize HB donors and acceptors inside DNA bases [181]. Below, to confront this opinion, we propose an analytical model for protein-DNA ES recognition. Afterwards, via a systematic computational analysis of PBD structures of protein-DNA complexes, we justify this model for large architectural complexes of pro- and eu-karyotes.

Model of ES recognition. A simple 1D model of DNA-protein recognition based on complementarity of their charge patterns was proposed in Ref. [182]. DNA and protein charge lattices were set commensurate for the cognate site and de-correlated for the rest of the DNA, Fig. 20. In the model, some random charge displacement fields along DNA Δ_n and protein δ_m mimic a sequence specificity of their charge patterns. This idealized protein is attracted stronger to a particular hom/matching segment on the idealized DNA.

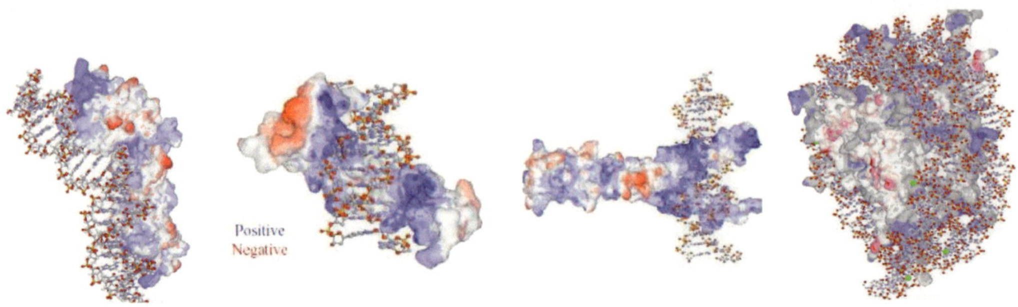

Fig. 19. The distribution of ES potential on DNA-protein complexes of specifically bound lac repressor 1l1m.pdb, zinc finger ZIF268 1aay.pdb, leucine zipper GCN4 1ysa.pdb, and 146 bp NCP 1aoi.pdb (from left to right). The structures are visualized by MDL Chime and Protein Explorer programs, using the PDB files of the complexes. Images are not to scale.

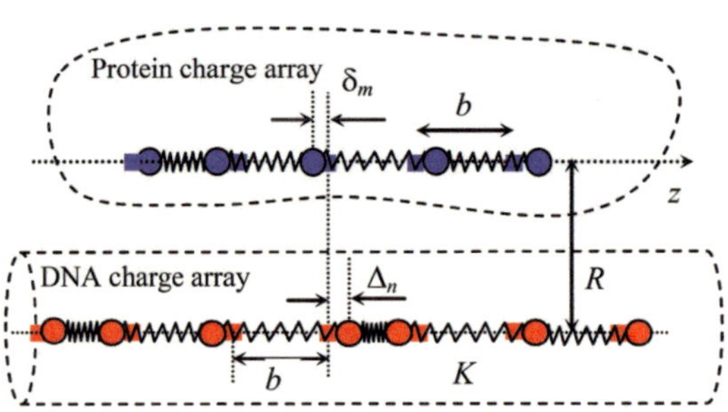

Fig. 20. Schematic 1D-model of "protein-DNA" ES recognition. Charge positions on DNA and protein vary in a random fashion about quasi-periodic positions on 1D lattice.

The average ES recognition energy of a protein to this target site has been derived in the linear PB theory, for random realization of Δ and δ fields. Both long- and short-range order situations for the charges on DNA were investigated (only long-range order results are shown here). For the parameters typical for lac-repressor-like proteins (~10 positive charges at R=1nm from the DNA) the ES recognition well amounts to $\approx$3-10k_BT in depth and a couple of nm in width, Fig. 21. Thus, this short-range well cannot serve as "ES funnel" that would direct diffusing proteins from far away on the DNA to this charge-hom binding site. This ES well is thus not expected to facilitate strongly the protein diffusion on DNA, the phenomenon known to take place e.g. for lac- and gal-repressors.

The well depth scales linearly with the number of charges in the hom domain, M. Larger magnitudes of charge deviations from their quasi-periodic positions on the lattice, described by $\Omega^2 = \langle \Delta_n{}^2 \rangle + \langle \delta_m{}^2 \rangle$, also make the well deeper. At zero [salt], the well depth drops as $\propto 1/R^3$ with protein-DNA separation R, while in electrolyte the decay is exponential, $\propto \exp[-R/\lambda_D]$. For weak charge fluctuations in no-salt limit the model returns an elegant expression for the average ES recognition energy well

$$\Delta E(\Delta z) = -\frac{k_B T l_B M \Omega^2 \varepsilon}{2\varepsilon_c} \frac{R^2 - 2\Delta z^2}{\left(R^2 + \Delta z^2\right)^{5/2}} \tag{6}$$

Here Δz is the mutual protein-DNA sliding distance with respect to the position of complete DNA and protein homology overlap at Δz=0, see Fig. 21. The charges are assumed to interact through a weakly-polarizable low-dielectric medium between DNA and protein, with the dielectric constant of $\varepsilon_c = 2$.

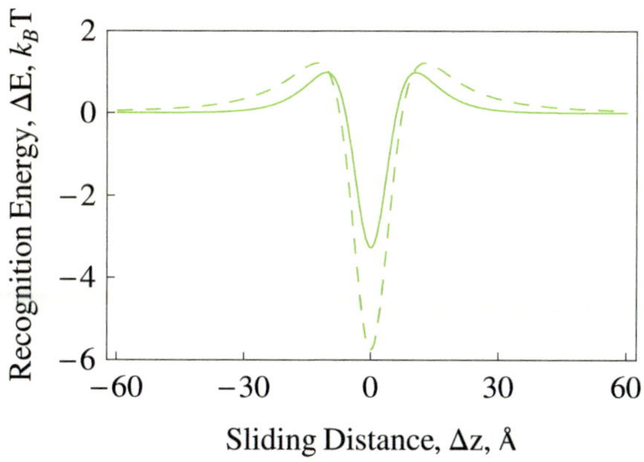

Fig. 21. ES recognition energy well upon sliding of a 1D "protein" over 1D DNA lattice. Parameters: M=11 charges in the hom region, R=10Å, $\Omega^2 = 2$ Å². The dashed curve is zero-salt limit, the solid curve describes a reduced ES recognition at $1/\kappa = 7$ Å.

In this 1D model, the recognition well is accompanied by *energetic barriers*, both for the exact solution and simplistic expression in Eq. 6. The barriers disappear when charge displacements perpendicular to the DNA-protein plane are also taken into account. A generalization of this 1D model for protein and DNA charge displacements for 2D and 3D is more realistic, but only computationally feasible. Also note that some specific, not fully random displacement fields Δ and δ, can mimic charge patterns on a particular DNA sequence and for a given protein. Protein-DNA ES recognition well then resembles DNA-DNA barrier-free ES hom recognition well, sketched in Fig. 16. The calculation of DNA-protein ES recognition is methodologically similar to that for DNA-DNA recognition funnel, Fig. 16.

Let us consider one physical implication of this ES recognition. We have calculated [182] that this well is capable to slow down the protein diffusion, provoking protein trapping for ~μs-ms near this hom site on DNA. This time is long enough to allow some conformational changes in the protein structure (domain motions, rotation of side-chains, allosteric transition, etc.). Various protein conformation being sampled might trigger a stronger (chemical or HB) protein binding to this particular DNA fragment.

Our hypothesis is thus a *two-step* mechanism of recognition for some proteins. First, a DNA-binding protein scans the ES surface of DNA for a charge-complementary site. In this "searching" mode, the protein structure is flexible and adaptable to the pattern of interaction sites on DNA. When a commensurate DNA fragment is found, some interaction-induced folding solidifies the protein structure, switching it into the "binding" mode that enables stronger and more specific contacts with the DNA. Cumulative ES and HB contacts rigidify the protein structure and give rise to formation of specifically bound DNA-protein complexes.

This kind of hot-spot two-step recognition mechanism is pretty common in structural molecular biology, both for protein-DNA and protein-protein complexes. For the latter, the two-step docking directs the assembly pathway into the native structure by "anchoring" of a shape-complementary relatively rigid "key" domain of one protein into a "lock" domain in the surface of another protein [183]. This process is accompanied by a large burial of solvent accessible surface area and this water release amplifies further docking of proteins into a tight complex.

Analysis of PDB structures. To justify the analytical predictions above, the detailed analysis of PDB structures of different classes of proteins in their complexes with the DNA has been performed. The distribution of NH^{2+} groups on Arginine$^+$ and Lysine$^+$ residues in DNA-binding domains of DNA-protein complexes has been examined [184]. In particular, large structural complexes were studied: the NCPs of eukaryotes and architectural proteins of prokaryotes, both involving extensive DNA wrapping around the basic protein cores and featuring mainly ES mode of binding. A home-written Mathematica 6 computer code was used for extracting the coordinates of DNA phosphates PO^{4-} and protein's N^+ and O^- atoms on the charged amino acids from the PDB files. We could analyse the distances from N+ atoms on Arg$^+$ and Lys$^+$ that are within ≈7Å from the closest (s_1) and next closest (s_2) DNA phosphates on the same DNA strand, see Fig. 22. The statistics of ES contacts and salt bridges in these DNA-protein complexes has thus been restored. Smaller cut-off distances of 3-5Å can also be used, to minimize the contributions of charges across the narrow DNA groove. Note that fluctuation-induced uncertainties in positions of protein charges in crystals of many DNA-protein complexes are often ~1-2Å. This is much smaller than the relevant periodicity in the system, the phosphate-phosphate separation along the DNA helical strand, s_{ph} ≈7Å.

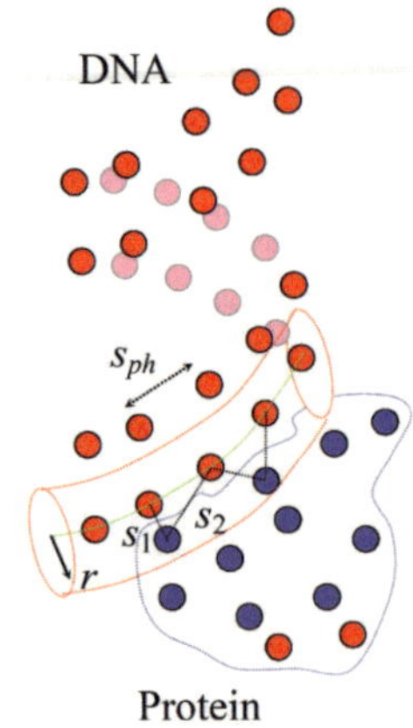

Fig. 22. Definition of $s_{1,2}$ distances for the protein positive charges (in blue) which are closer than $r \sim l_B \approx 7$Å to the negative DNA phosphates (red helix).

For the NCPs, the histone positive charges are mainly localized in the outer "ring" of the octamer, close to the wrapped DNA, see Fig. 19d, with Asp- and Glu- acids being rather inside the core, further away from the DNA. Our analysis demonstrates that N^+ atoms on Arg^+ and Lys^+ track the positions of *individual* DNA phosphates, as visualized in the histogram $s_1 - s_2$ for all N^+ charges in DNA vicinity, Fig. 23. The *bimodal distributions* were detected both for individual NCPs (a good statistics can be achieved for a single particle) and for the entire family of 146 bp long complexes [184]. This indicates that N^+ on Lys and Arg are encountered measurably more often close to one of the neighboring DNA phosphates than between the two, maximizing the ES attraction to DNA. As the structure and positions of DNA phosphates strongly correlate to DNA bp sequence [52], this detected "charge tracking" for DNA sequences wrapped in NCPs yields sequence-specific ES DNA-protein forces. This supports our model hypothesis above about commensurate charge lattices on DNA target sequence and on the protein. This fact can contribute to NCP positioning on genomic DNAs, interfering with the mechanism of sequence-specific DNA bendability believed to govern this process [185]. A similar bimodal distribution was obtained for prokaryotic NCP analogs (not shown) [184].

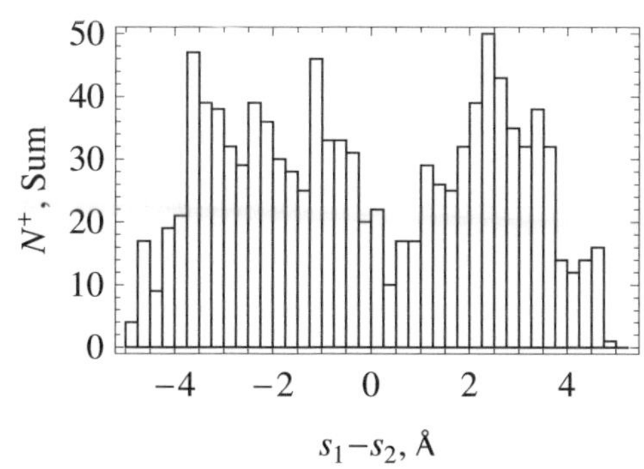

Fig. 23. A bimodal distribution of $s_1 - s_2$ distances for 14 NCP complexes. It indicates the ES recognition of individual DNA phosphates by the closest N^+ atoms on Arg and Lys basic residues of the histone core proteins.

The specificity of ES binding of Arg and Lys of core histones into the DNA grooves on NCPs has recently been examined by other groups too [186]. The ES-directed localization of Arg+ in the minor grooves in AT-rich DNA regions was confirmed for NCPs in Ref. [187]. These studies emphasize that for DNA sequences wrapped in NCPs the AT-tracts have particularly narrow minor grooves that offers "attractive" sites for Arg^+ and Lys^+ binding every time the DNA minor groove faces the histone octamer. Arg was claimed to be preferred over Lys in the DNA minor grooves because of a lower self-energy cost to remove a larger guanidinium group of Arg+ from its hydrated state in solution. The reason is the Born ES self-energy that scales inversely proportional with the "ion" radius.

Being valid for large structural complexes, the ES complementarity model fails for small DNA-protein complexes, with the standard simple motifs of DNA recognition (e.g., helix-turn-helix, zinc finger, leucine zipper). For a large set of small proteins from these families, we could not detect any statistical preference in distribution of Lys and Arg close to DNA phosphates [184]. To make a more definite conclusion, some redundancy in protein structures is to be excluded and a grouping into smaller, more specific protein sub-families is to be performed.

A tentative explanation is however as follows. For large complexes, with ~30-100 ES DNA-protein contacts, the ES energy gain being ES-commensurate can reach ~10-30k_BT and proteins appear to utilize it for sequence-specific binding to DNA. For small proteins, with only ~3-10 ES contacts and much weaker ES binding, other interactions (such as HBs) are likely to direct the recognition of specific DNA sequences in complexes.

Note that DNA-protein ES commensurability for NCPs and their prokaryotic analogs resembles a *zipper-like* positioning of positive and negative amino acids along interfaces of many protein-protein complexes [18,188]. For the latter, despite hydrophobic residues often dominate the overall binding affinity, these non-charged amino acids might be too abundant to ensure a proper degree of the binding specificity. The latter might stem form *charge patchiness* and propensity of HB formation between the residues along the contact surface of the bound proteins [189]. Analogously, for DNA-DNA interactions, overviewed in Sec. 2, we have seen that DNA-DNA attraction is only possible for entirely commensurate or hom sequences, for which *charge zipper motif* is realized. All these similarities for protein-DNA, protein-protein, and DNA-DNA interactions reflect different aspects of the universal principle of ES complementarity in structural molecular biology.

Challenges and Perspectives. Several issues of description of biophysics of DNA-protein interactions challenge future theoretical developments. First, the computational analysis of PDB data presented above provides us only with a *statistical preference* of distribution of protein charges in DNA proximity. To evaluate the ES binding energies of DNA-protein complexes, one needs to speculate about the value of the dielectric constant ε_c in spaces between DNA and protein. Because of dielectric saturation effects in confined/hydrated water molecules on the charged objects [190], its value can vary widely, ε_c ~2÷30 [11,12]. So does the ES interaction energy [191]. Another important ES issue is the charged state of ionizable protein residues in a particular neighborhood in DNA-protein complexes (e.g., Hist), with their pK_a value being affected by local ES potential, geometrical shape of the protein surface, [salt], local dielectric permittivity, etc. [11,192].

Non-ES van der Waals and HB contacts, as well as the entropic terms associated with water release and counterion evaporation upon protein-DNA binding, are to be quantified in the future models as well. To make unambiguous conclusion about the mechanism of binding specificity for a given DNA-protein complex, the ES preference of Arg/Lys positioning with respect to DNA phosphates has to be supplemented by the analysis of HB formation propensity between the protein residues and DNA bases [181]. Also, one has to keep in mind that the protein (and DNA) structure visualized by x-rays in crystals exposed to special crystallization buffers [193] might measurably differ from real structures stable at physiological conditions.

There exists an opinion in the literature that ES contacts of charged residues in protein-DNA complexes [4] and along protein-protein interfaces [11] might (somewhat counter-intuitively) *destabilize instead of stabilize* their binding. The argument goes as follows. The release of ES-profitable structured water shells around the constituents often accompanies

DNA-protein complex formation [4,194]. And, it is possible that protein and DNA charged groups complexed together via ES attraction do not fully compensate for the energetic loses upon their "ES desolvation". The latter depends crucially on the ε-value assigned to the protein and its immediate vicinity. For DNA-protein complexes, the entropic effects of condensed cations released from the DNA, with the number defined by the slope of log[binding constant] on log [salt], are often presented as the main *driving force* for the complexation. Here, the situation is rather similar to counterion release from DNA-(CL membrane) complexes, Sec. 4. In both cases, we however tend to think that the direct ES attraction between the oppositely charged components of the system governs/directs the complex formation, while the entropic free energy gain due to the release of condensed counterions accompanies this process.

11. Conclusions and outlook

In this chapter, we focused on recent developments and new viewpoints on ES effects for a number of biological DNA-related systems. Several experimental achievements and DNA-related phenomena discovered in the last years have been overviewed, which challenge both theoretical and computational modeling. Some analytical insights from our recent studies are discussed, which uncover general principles behind charge-mediated DNA-DNA, NCP-NCP, and DNA-protein interactions. We aimed at describing macroscopic effects having their possible origin in ES interactions as well as at trying to establish correlations between the structure of the system components and their function. The advanced theoretical and computational approaches developed in our studies on DNA-DNA, DNA-membrane and DNA-proteins interactions can find their applications in bio-technology and nano-engineering.

The PE models for DNA and available structure information for the proteins have been applied to some nano-technology applications, the principles of bio-molecular DNA-protein recognition, and self-assembly. Despite inherent limitations of the mean-field PB-like theories applied to the DNA, the approaches developed often enabled us to rationalize the structural properties of the system as dictated by intermolecular forces. The conceptual framework proposed in the chapter allows us to anticipate the physical effects in these DNA-related systems that are still too large for modern *ab initio* computer simulations. Clearly, more work is to be done to achieve a quantitative understanding of these complex phenomena. In particular, the physical properties of inter-connected NCPs in 30 nm chromatin fibers and DNA packaging inside bacteriophages feature a number of important biological details to be incorporated in future theoretical models. Another area is ES effects in protein-mediated loop formation in DNA [195], DNA plectonemes [196,197] and cyclization [198], as well as DNA wrapping in NCPs [199]. These interesting phenomena are however beyond the scope of this contribution.

One hot and intriguing domain of our ES-related biological research is DNA packaging inside viral capsids and self-assembly of viral shells from the capsid proteins [27]. Both processes are highly sensitive to salt conditions that control protein-protein and DNA-DNA ES forces. We argue here that the accurate physical description of DNA compactification inside viral shells demands the application of all theories and models presented in the main text. Let us list the effects one by one.

Many ds-DNA bacteriophages pack their DNA in a very dense and well-organized fashion [200,201,202]. DNA densities can reach $R \approx 23$-28 Å between DNA axes, creating osmotic pressures of up to ~50 atm inside the shells [203,204,205]. At such DNA densities, the effects of DNA helical structure onto DNA-DNA ES forces are going to be extremely pronounced, see

Fig. 24. These effects have been however largely neglected in the existing theories of DNA compaction in ds DNA viruses.

As the capsids of some ds-DNA viruses are penetrable for small ions [40], the presence of di- and tri-valent cations in the solution can render DNA-DNA ES forces inside the capsids more attractive [212]. This will ease DNA packaging into and inhibit the DNA ejection from such capsids. Indeed, only 1 mM of spermine^{4+} in the buffer blocks nearly 90% of DNA inside the λ-phage capsids [206]. The counterion-mediated DNA-DNA attraction inside viral capsids is a clear target for our ES interaction theory presented in Sec. 2.

Typical for ds-DNA viruses are the concentric rings of DNA [207,208,209], with the DNA layers that are closer to the viral shell being resolved better by the cryo-EM image reconstruction, see Fig. 24. This corresponds to a coaxial inverse-spool model of DNA packing, with the outer (more ordered) shells of the DNA spool being filled first. Recently, oriented DNA toroids condensed with spermine^{4+} inside T4 phages [75] and DNA "domain-wall" transitions upon DNA ejection from T5 phages [210] have been clearly visualized by cryo-EM. The interaction of toroidal DNA condensates overviewed in Sec. 3 with the confining protein shells of the capsid is a proper model to describe this "deformed toroid" conformation of spooling DNA [76]. Cholesteric ES effects, see Sec. 5, onto the DNA packing properties were also argued to be important for many ds DNA viruses. Lastly, the pauses in DNA packaging and ejection caused by necessary rearrangements of DNA spool [211] might stem from ES friction between the densely packed DNA layers inside the capsid, see Sec. 8.

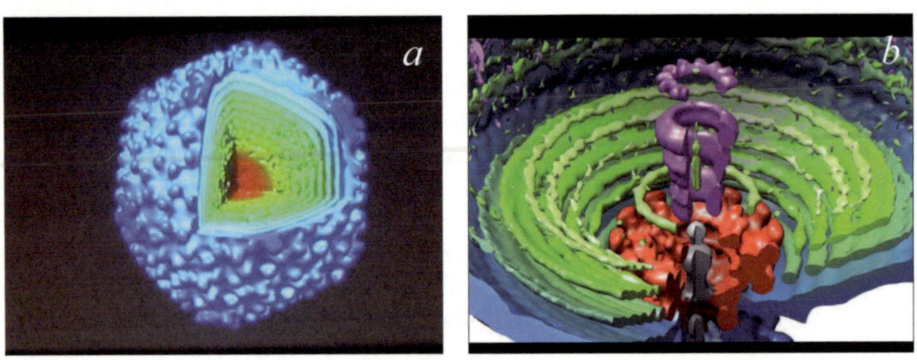

Fig. 24. The result of 3D reconstruction of cryo-EM images with DNA layers inside P22 virus (a), with DNA layers being more ordered near the portal region (b, bottom). Images are the courtesy of J. Johnson.

12. Acknowledgment

I am thankful to T. Bellini, R. Everaers, N. Kleckner, A. Kolomeisky, A. Kornyshev, D. Lee, S. Leikin, S. Malinin, W. Olson, E. Petrov, A. Poghossian, M. Prentiss, E. Starostin, R. Winkler,

G. Wuite and G. Zanchetta for many stimulating discussions and scientific correspondence. Without a constant help and encouragement of my collaborators (see joint publications below) this chapter would not be possible. A part of this work was supported by the German Research Foundation, DFG Grants CH 707/5-1 and CH 707/2-2. Because of space restrictions, only a small fraction of relevant studies has been cited in the text (my apologies to the authors which could not be mentioned).

13. References

[1] W. M. Gelbart et al., Phys. Today, 53 38 (2000).
[2] S. McLaughlin, Annu. Rev. Biophys. Biophys. Chern 18 113 (1989).
[3] C. G. Kalodimos et al., Science, 305 386 (2004).
[4] B. Honig and A. Nicholls, Science, 268 1144 (1995).
[5] K. Luger et al., Nature, 389 251 (1997).
[6] T. J. Richmond and C. Davey, Nature, 423 145 (2003).
[7] J. H. Morais-Cabral et al., Nature, 414 37 (2001).
[8] G. C. L. Wong and L. Pollack, Annu. Rev. Phys. Chem., 61 171 (2010).
[9] A. A. Kornyshev et al., Rev. Mod. Phys., 79 943 (2007).
[10] S. J. Chen, Ann. Rev. Biophys., 37 197 (2008).
[11] A. Warshel and T. S. Russell, Quart. Rev. Biophys., 17 283 (1984).
[12] F. B. Sheinerman et al., Curr. Opin. Struct. Biol., 10 153 (2000).
[13] A. Warshel et al., Biochim. Biophys. Acta, 1764 1647 (2006).
[14] R. P. Rand, Ann. Rev. Biophys. Bioeng., 10 277 (1981).
[15] G. Cevc, Biochim. Biophys. Acta, 1031 311 (1990).
[16] R. Messina, J. Phys.: Cond. Matter, 21 113102 (2009).
[17] P. H. von Hippel, Science, 305 350 (2004).
[18] A. J. McCoy et al., J. Mol. Biol., 268 570 (1997).
[19] H. Schiessel, J. Phys.: Cond. Matt., 15 R699 (2003).
[20] C. L. Woodcock and R. P. Ghosh, Cold Spring Harb. Perspect. Biol., 2 a000596 (2010).
[21] C. M. Knobler and W. M. Gelbart, Ann. Rev. Phys. Chem., 60 367 (2009).
[22] D. E. Draper, Biophys. J., 95 5489 (2008).
[23] N. A. Baker et al., PNAS, 98 10037 (2001).
[24] R. D. Kornberg et al., Science, 292 1863 (2001).
[25] V. A. Parsegian and D. Gingell, J. Adhesion, 4 283 (1972).
[26] A. A. Kornyshev and S. Leikin, J. Chem. Phys., 107 3656 (1997).
[27] A. G. Cherstvy, Phys. Chem. Chem. Phys., 13 9942 (2011).
[28] G. S. Manning, Q. Rev. Biophys., 11 179 (1978).
[29] R. M. M. Smeets, Nano Lett., 6 89 (2006).
[30] S. van Dorp et al., Nature Physics, 5 347 (2009).
[31] A. G. Cherstvy, J. Chem. Phys., 123 116101 (2005).
[32] J. DeRouchey et al., Biophys. J. 99 2608 (2010).
[33] D. C. Rau and V. A. Parsegian, Science, 249 1278 (1990).
[34] S. Leikin et al., PNAS, 91 276 (1994).
[35] V. A. Parsegian et al., PNAS, 76 2750 (1979).

[36] D. C. Rau and V. A. Parsegian, Biophys. J., 61 260 (1992); ibid., 61 272 (1992).
[37] B. A. Todd et al., Biophys. J., 94 4775 (2008).
[38] H. H. Strey et al., Phys. Rev. E, 59 999 (1999).
[39] P. Shah and E. Swiatlo, Molec. Microbiol., 68 4 (2008).
[40] B. N. Ames and D. T. Dubin, J. Biol. Chem., 235 769 (1960).
[41] M. Kanduč et al., Soft Matter, 5 868 (2009).
[42] V. B. Teif and K. Bohinc, Progr. Biophys. Mol. Biol., 105 208 (2011).
[43] I. Rouzina and V. A. Bloomfield, J. Phys. Chem., 100 9977 (1996).
[44] H. Boroudjerdi et al., Phys. Rep., 416 129 (2005).
[45] Y. Levin, Rep. Prog. Phys., 65 1577 (2002).
[46] N. Gronbech-Jensen et al., Phys. Rev. Lett., 78, 2477 (1997).
[47] E. Allahyarov et al., Phys. Rev. E, 69 041904 (2004).
[48] M. Deserno et al., Macromol., 36 249 (2003).
[49] B. Luan and A. Aksimentiev, JACS, 130 15754 (2008).
[50] A. A. Kornyshev and S. Leikin, Phys. Rev. Lett., 82 4138 (1999).
[51] A. A. Kornyshev and S. Leikin, PNAS, 95 13579 (1998).
[52] W. K. Olson et al., PNAS, 95 11163 (1998).
[53] A. A. Kornyshev and S. Leikin, Phys. Rev. Lett., 86 3666 (2001).
[54] A. G. Cherstvy et al., J. Phys. Chem. B, 106 13362 (2002).
[55] A. G. Cherstvy et al., J. Phys. Chem. B, 108 6508 (2004).
[56] A. A. Kornyshev and A. Wynveen, Phys. Rev. E, 69 041905 (2004).
[57] A. Wynveen et al., Nucl. Acid Res., 36 5540 (2008).
[58] A. A. Kornyshev and S. Leikin, Phys. Rev. Lett., 84 2537 (2000).
[59] A. A. Kornyshev et al., Eur. Phys. J. E, 7 83 (2002).
[60] A. Wynveen et al., Eur. Phys. J. E, 16 303 (2005).
[61] H. M. Harreis et al., Phys. Rev. Lett., 89 018303 (2002).
[62] A. A. Kornyshev et al., Phys. Rev. Lett., 95 148102 (2005).
[63] H. H. Strey et al., Phys. Rev. Lett., 84 3105 (2000).
[64] D. J. Lee et al., J. Phys. Condens. Matter, 22 072202 (2010).
[65] D. J. Lee et al., J. Phys. Chem. B, 114 11668 (2010).
[66] D. J. Lee, J. Phys. Condens. Matter, 22 414101 (2010).
[67] A. A. Kornyshev, Phys. Chem. Chem. Phys., 12 12352 (2010).
[68] H. H. Strey et al., Phys. Rev. Lett., 78 895 (1997).
[69] G. M. Grason, EPL, 83 58003 (2008).
[70] A. G. Cherstvy, PhD Thesis, Düsseldorf University, 2002.
[71] V. A. Bloomfield, Curr. Opin. Struct. Biol., 6 334 (1996).
[72] S. Levin-Zaidman, Science, 299 254 (2003).
[73] J. Englander et al., J. Bacteriol., 186 5973 (2004).
[74] N. V. Hud et al., Biochem. Biophys. Res. Comm., 193 1347 (1993).
[75] A. Leforestier and F. Livolant, PNAS, 106 9157 (2009).
[76] A. Leforestier et al., Biophys. J., 100 2209 (2011).
[77] N. V. Hud and K. N. Downing, PNAS, 98 14925 (2001).
[78] S. M. Douglas et al., Nature, 459 414 (2009).
[79] H. Dietz et al., Science, 325 725 (2009).

[80] A. G. Cherstvy, J. Phys.: Cond. Matter, 17 1363 (2005).
[81] C. G. Baumann et al., PNAS, 94 6185 (1997).
[82] J. P. Peters and L. J. Maher III, Quart. Rev. Biophys., 43 23 (2010).
[83] J. Ubbink and T. Odijk, EPL, 33 353 (1996).
[84] J. Kindt et al., PNAS, 98 13671 (2001).
[85] I. M. Kulic et al., EPL, 67 418 (2004).
[86] N. V. Hud et al., PNAS, 92 3581 (1995).
[87] B. van den Broek et al., Biophys. J., 98 1902 (2010).
[88] C. Battle et al., Phys. Rev. E, 80 031917 (2009).
[89] Y. Sh. Mamasakhlisov et al., Phys. Rev. E, 80 031915 (2009).
[90] G. G. Pereira and D. R. M. Williams, Biophys. J., 80 161 (2001).
[91] C. Herold et al., Phys. Rev. Lett., 104 148102 (2010).
[92] E. Petrov and A. G. Cherstvy, work in preparation.
[93] J. O. Raedler et al., Science, 275 810 (1997).
[94] G. C. L. Wong et al., Science, 288 2035 (2000).
[95] U. Raviv et al., PNAS, 102 11167 (2005).
[96] L. Tang et al., Nature Materials, 3 615 (2004).
[97] G. Martini and L. Ciani, Phys. Chem. Chem. Phys., 11 211 (2009).
[98] K. K. Ewert et al., Expert Opin. Biol. Therapy, 55 33 (2005).
[99] I. Koltover et al., Science, 281 78 (1998).
[100] G. Caracciolo et al., J. Phys. Chem. B, 114 2028 (2010).
[101] K. Wagner et al., Langmuir, 16 303 (2000).
[102] D. Harries et al., Biophys. J., 75 159 (1998).
[103] S. May et al., Biophys. J., 78 1681 (2000).
[104] D. Harries et al., Coll. Surf. A, 208 41 (2002).
[105] O. Farago et al., Phys. Rev. Lett., 96 018102 (2006).
[106] L. Gao et al., J. Phys. Chem. B, 114 7261 (2010).
[107] G. Tresset and Y. Lansac, J. Phys. Chem. Lett., 2 41 (2011).
[108] C. Fleck et al., Biophys J., 82 76 (2002).
[109] H. Schiessel and H. Aranda-Espinoza, Eur. Phys. J. E, 5 499 (2001).
[110] A. G. Cherstvy, J. Phys. Chem. B, 111 12933 (2007).
[111] T. Salditt et al., Phys. Rev. E, 58 889 (1998).
[112] R. Menes et al., Europ. Phys. J. E, 1 345 (2000).
[113] I. Koltover et al., PNAS, 97 14046 (2000).
[114] E. Raspaud et al., Biophys. J., 88 392 (2005).
[115] Y. S. Tarahovsky et al., Biophys. J., 87 1054 (2004).
[116] A. Leforestier and F. Livolant, Biophys. J., 65 56 (1993).
[117] A. D. Rey, Soft Matter, 6 3402 (2010).
[118] E. Grelet and S. Fraden, Phys. Rev. Lett., 90 198302 (2003).
[119] L. Rudd et al., Phys. Chem. Chem. Phys., 8 4347 (2006).
[120] J. Lepault et al., EMBO J., 6 1507 (1987).
[121] F. Livolant and M. F. Maestre, Biochem. 27 3056 (1988).
[122] N. S. Blanc et al., J. Struct. Biol., 134 76 (2001).
[123] C. B. Stanley et al., Biophys. J., 89 2552 (2005).

[124] J. Pelta et al., Biophys. J., 71 48 (1996).
[125] A. A. Kornyshev and S. Leikin, Phys. Rev. E, 62 2576 (2000).
[126] J. P. Straley, Phys. Rev. A, 14 1835 (1976).
[127] F. Tombolato and A. Ferrarini, J. Chem. Phys., 122 054908 (2005).
[128] A. G. Cherstvy, J. Phys. Chem. B, 112 12585 (2008).
[129] F. G. Donnan, Chem. Rev., 1 73 (1924).
[130] G. Zanchetta et al., PNAS, 107 17497 (2010).
[131] M. Nakata et al., Science, 318 1276 (2007).
[132] G. Zanchetta et al., PNAS, 105 1111 (2008).
[133] P. Varnai and Y. Timsit, Nucl. Acid Res., 38 4163 (2010).
[134] E. Frezza et al., Soft Matter, 7 9291 (2011).
[135] A. Goldar et al., J. Phys. Cond. Matter, 20 035102 (2008).
[136] C. L. Woodcock et al., Chromosome Res., 14 17 (2006).
[137] J. T. Finch and A. Klug, PNAS, 73 1897 (1976).
[138] T. Schalch et al., Nature, 436 138 (2005).
[139] P. J. J. Robinson et al, PNAS, 103 6506 (2006).
[140] K. Luger and T. J. Richmond, Curr. Opin. Struct. Biol., 8 33 (1998).
[141] C. L. White et al., EMBO J., 20 5207 (2001).
[142] Y. Cui and C. Bustamante, PNAS, 97 127 (2000).
[143] M. Kruithof et al., Nat. Struct. Mol. Biol., 16 534 (2009).
[144] S. Mangenot et al., Biophys. J., 84 2570 (2003).
[145] A. Leforestier et al., Biophys. J., 81 2414 (2001).
[146] K. Luger et al., PNAS, 101 6864 (2004).
[147] A. G. Cherstvy and R. Everaers, J. Phys. Cond. Mat., 18 11429 (2006).
[148] A. A. Kornyshev and A. Wynveen, PNAS, 106 4683 (2009).
[149] A. Barzel and M. Kupiec, Nature Rev. Gen., 9 27 (2008).
[150] A. Weiner et al., Nature Rev. Microbiol., 7 748 (2009).
[151] W. K. Olson and V. B. Zhurkin, Curr. Opin. Struct. Biol., 10 286 (2000).
[152] A. A. Gorin et al., J. Mol. Biol., 247 34 (1995).
[153] B. S. Singer et al., Cell, 31 25 (1982).
[154] B. M. Weiner and N. Kleckner, Cell, 77 977 (1994).
[155] S. Inoue et al., Biochemistry, 46 164 (2007).
[156] G. S. Baldwin et al., J. Phys. Chem B, 112 1060 (2008).
[157] M. Prentiss et al., PNAS, 106 19824 (2009).
[158] N. Kleckner, PNAS, 93 8167 (1996).
[159] A. G. Cherstvy, J. Mol. Recogn., 24 283 (2011).
[160] M. Urbakh and E. Meyer, Nature Mat., 9 8 (2010).
[161] I. Barel et al., Phys. Rev. Lett., 104 066104 (2010).
[162] N. Clauvelin et al., Biophys. J., 96 3716 (2009).
[163] R. T. Dame et al., Nature, 444 387 (2006).
[164] M. C. Noom et al., Nature Methods, 4 1031 (2007).
[165] A. G. Cherstvy, J. Phys. Chem. B, 113 5350 (2009).
[166] A. G. Cherstvy and R. G. Winkler, J. Chem. Phys., 120 9394 (2004).

[167] D. Poland and H.A. Scheraga, Theory of Helix-Coil Transition in Biopolymers, Academic Press, New York (1970).
[168] M. Peyrard and A. R. Bishop, Phys. Rev. Lett., 62 2755 (1989).
[169] D. Jost and R. Everaers, Biophys. J., 96 1056 (2009).
[170] A. G. Cherstvy and A. A. Kornyshev, J. Phys. Chem. B, 109 13024 (2005).
[171] D. Grasso et al., Liquid Cryst., 9 299 (1991).
[172] R. Jin et al., JACS, 125 1643 (2003).
[173] H. Long et al., J. Phys. Chem. B, 110 2918 (2006).
[174] O.-S. Lee et al., J. Phys. Chem. Lett., 1 1781 (2010).
[175] C. Schildkraut and S. Lifson, Biopolym., 3 195 (1965).
[176] M. Peyrard et al., Phys. Rev. E, 83 061923 (2011).
[177] A. Poghossian et al., Sens. & Actuators B, 111-112 470 (2005).
[178] A. Poghossian et al., submitted, (2011).
[179] C. G. Kalodimos, et al., Science, 305 386 (2004).
[180] P. H. von Hippel and O. G. Berg, PNAS, 83 1608 (1986).
[181] N. M. Luscombe et al., NAR, 29 2860 (2001).
[182] A. G. Cherstvy et al., J. Phys. Chem. B, 112 4741 (2008).
[183] D. Rajamani et al., PNAS, 101 11287 (2004).
[184] A. G. Cherstvy, J. Phys. Chem. B, 113 4242 (2009).
[185] E. Segal et al., Nature, 442 772 (2006).
[186] D. Wang et al., J. Biomol. Struct. Dyn., 27 843 (2010).
[187] B. Honig et al., Nature, 461 1248 (2009).
[188] T. Takahashi, Adv. Biophys., 34 41 (1997).
[189] O. Keksin et al., J. Mol. Biol., 245 1281 (2005).
[190] C. N. Schutz and A. Warshel, Proteins, 44 400 (2001).
[191] A. T. Fenley et al., J. Chem. Phys., 129 075101 (2008).
[192] J. Antosiewicz et al., Biochem., 35 7819 (1996).
[193] R. E. Dickerson et al., PNAS, 91 3579 (1994).
[194] V. K. Misra et al., Biophys. J., 75 2262 (1998).
[195] A. G. Cherstvy, Europ. Biophys. J., 40 69 (2011).
[196] A. G. Cherstvy, J. Biol. Phys., 37 227 (2011).
[197] R. Cortini et al., Biophys. J, 101 875 (2011).
[198] A. G. Cherstvy, J. Phys. Chem. B, 115 4286 (2011).
[199] A. G. Cherstvy and R. G. Winkler, J. Phys. Chem. B, 109 2962 (2005).
[200] W. C. Earnshaw and S. C. Harrison, Nature, 268 598 (1977).
[201] W. M. Gelbart and C. M. Knobler, Physics Today, 61 42 (2008).
[202] W. M. Gelbart and C. M. Knobler, Science, 323 1682 (2009).
[203] D. E. Smith et al., Nature, 413 748 (2001).
[204] D. N. Fuller et al., PNAS, 104 11245 (2007).
[205] P. Grayson et al., PNAS, 104 14652 (2007).
[206] A. Evilevitch et al., Biophys. J., 94 1110 (2008).
[207] M. E. Cerritelli et al., Cell, 91 271 (1997).
[208] A. Fokine et al., PNAS, 101 6003 (2004).
[209] J. Johnson et al., Science, 312 1791 (2006).

[210] A. Leforestier and F. Livolant, J. Mol. Biol., 396 384 (2010).
[211] D. Marenduzzo et al., PNAS, 106 22269 (2009).
[212] A. Siber et al., Phys. Chem. Chem. Phys., DOI: 10.1039/C1CP22756D

Part 2

Bioengineering

2

Electrostatics in Protein Engineering and Design

I. John Khan, James A. Stapleton, Douglas Pike and Vikas Nanda

University of Medicine and Dentistry of New Jersey, Piscataway, NJ, USA

1. Introduction

The electrostatic interactions between charged atoms in natural proteins play a central role in specifying protein topology, modulating stability of the molecule, and allowing for the impressive catalytic properties of enzymes. In this chapter, we discuss how protein engineers use the principles of electrostatics and computational protein modeling to develop new proteins for biomedical and biotechnological applications. First, a general introduction is given to familiarize the reader with the important factors to consider in protein electrostatics, and the nature of these electrostatic forces. The next section describes various levels of theory used for modeling electrostatics in proteins. The last sections focus on specific applications in two conceptual classes: the engineering of ionic interactions (1) on protein surfaces, and (2) within the hydrophobic protein core. In both cases, the aim is to promote stability or to control molecular recognition.

2. Important factors influencing protein electrostatics

Interacting ionic species undergo rearrangement of their charge distributions under the influence of each other and their local environment. In electrostatics, we consider the static electrical field that is formed between these charged species once charge rearrangement has occurred. In the context of a protein, this amounts to looking at the many interactions among the polar and/or charged residues scattered throughout the three dimensional structure. Uncharged polar residues can form hydrogen bonding interactions with the hydroxyl (serine and threonine) and amide (asparagine and glutamine) hydrogen bond donors and acceptors on their side chains. Ionizable, or charged, residues have the following titratable side groups: carboxyl (asparate and glutamate), sulfhydryl (cysteine), hydroxyl (tyrosine), guanidino (arginine), amino (lysine), and imidazole (histidine). The ionization state of a titratable residue depends on its pK_a value or proton affinity, which represents the pH at which there is equilibrium between the neutral and charged forms of their respective functional groups.

Electrostatic interactions with the local environment influence the pK_a values of titratable residues. These factors are manifested in the following relationship for the pK_a of a buried residue (Bashford and Karplus 1990; Kaushik *et al.* 2006):

$$pK_a = pK_{a,model} + \Delta pK_{desolv} + \Delta pK_{back} + \Delta pK_{coulomb} \quad (1)$$

The pK_a value for each model residue ($pK_{a,model}$) has been experimentally determined. This value represents the pK_a of the residue when it is completely surrounded by water. Adjustments to the $pK_{a,model}$ as the residue becomes buried are due to the following three factors. The first is the ΔpK_{desolv}, which is the change in pK_a due to the unfavorable removal of an ionized residue from water to the hydrophobic protein core (i.e., a desolvation penalty). The second is the ΔpK_{back}, which is the change in pK_a due to interactions of the buried ionized residue with background charges present within the protein. Background charges are defined as the partial charges of atoms that are manifested as either permanent or induced dipoles in molecules. Examples of background charges are the permanent dipole of a water molecule or the permanent dipole that is formed by an α–helical domain of the protein. The third is the $\Delta pK_{coulomb}$, which is the change in pK_a due to charge-charge interactions among buried ionized residues (or with metal ions, if present). The signs of these ΔpK values will depend on whether the ionized residue is positive or negative, and on the strength of the electrostatic interactions.

The dominant forces to consider in protein electrostatics are the ion-ion, hydrogen bonding, ion-permanent dipole, and permanent dipole-permanent dipole interactions. The strength of these interactions are distance-dependent, as shown in Table 1, with the force of ion-ion pairing being exerted over a significantly longer range compared to weak non-electrostatic forces. For example, the electrostatic force between two charged residues Lys^+ and Glu^- decreases over a distance as $1/r$, whereas the van der Waals attraction between uncharged atoms decreases over a distance as $1/r^6$, where r is on the order of atomic distance. The attraction between the oppositely charged residues, such as Lys^+ and Glu^-, forms a salt bridge, where by definition, both the centroids of their side groups and the charged atoms lie within a range of 4-8 Å (Kumar and Nussinov 2002). Salt bridges, hydrogen bonding, and background charges are commonplace in protein structures, yet proteins are stabilized not only by these electrostatic forces but also by non-electrostatic interactions as well. Examples of non-electrostatic interactions are hydrophobic interactions, van der Waals interactions, disulfide bridges, or covalent bonds.

Type of interaction		Example	Distance dependence
Ionic	Electrostatic	Lys^+ --- Glu^- (salt bridge)	$1/r$
Hydrogen bonding	Electrostatic	Ser --- carboxyl of peptide bond	Bond length ~ 2.7 Å
Ionic/dipole	Electrostatic	Asp^- --- H_2O	$1/r^2$
Dipole/dipole (permanent)	Electrostatic	Helix dipole --- helix dipole	$1/r^3$
Dipole/dipole (induced)	Non-electrostatic	Dispersion forces	$1/r^6$

Table 1. Relative range of electrostatic and non-electrostatic interactions in proteins

Two other important factors influencing electrostatics in proteins are (1) the dielectric properties of the protein and its surrounding aqueous environment, and (2) the ionic strength of the aqueous environment. The dielectric coefficient (ε) is an indication of polarizability - how readily dipoles can reorient within the medium. In an aqueous environment, the dipoles of water molecules are free to reorient, hence the dielectric coefficient of water is relatively high with a value of ~78. In contrast, the dielectric coefficient of the protein is lower due to the limited mobility of the protein chain and the nonpolar nature of many amino acid residues. The dielectric coefficient within a protein varies with location, with values of 2-4 for regions having residues that are virtually inaccessible to water (i.e., the hydrophobic core), increasing to values of ~37 near the surface of the protein (Anslyn and Dougherty 2006). As a rule of thumb, we consider the range of electrostatic interactions to be dependent on the dielectric property of the medium according to the plots shown in Figure 1. For example, the energy between point charges in water (ε = 78) cannot be discriminated from baseline thermal energy at a separation of ~2 Å as a result of the charge screening by dipoles of water molecules. In the region below the protein surface (ε = 10) the effective separation increases to ~ 14Å, and within the hydrophobic core (ε = 4) the effective range can be greater than 30 Å. This difference implies that polar and charged residues have greater electrostatic potential when they are buried within the protein. The other factor influencing electrostatic interactions is the ionic strength which also has a screening effect of charge, particularly at the surface of the protein.

Here we focus on the treatment of electrostatics in protein engineering design. For a more general discussion of electrostatics in proteins, we refer the reader to several excellent reviews (Neves-Petersen and Petersen 2003; Bosshard *et al.* 2004; Jelesarov and Karshikoff 2009; Pace *et al.* 2009; Kukic and Nielsen 2010).

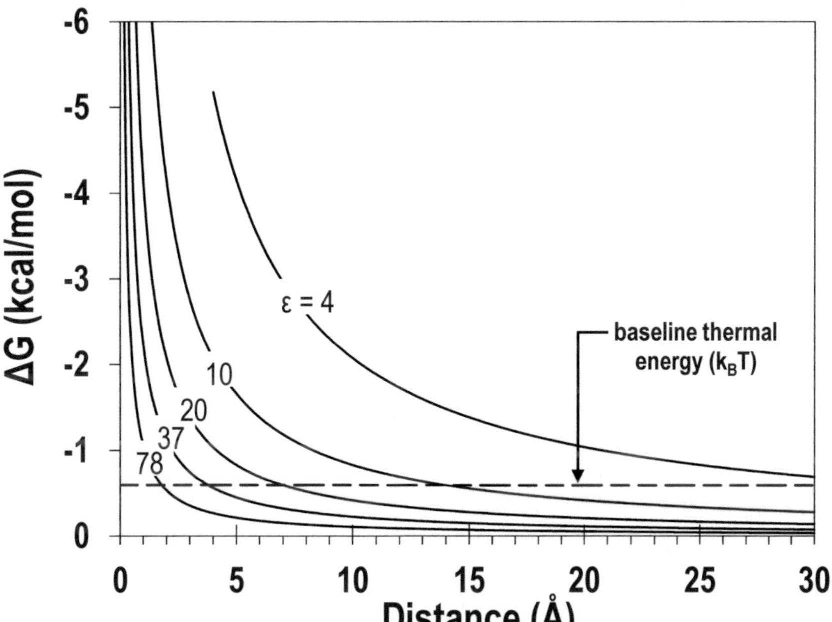

Fig. 1. The change in free energy (ΔG) associated with separating two point charges of opposite sign when surrounded by a medium of dielectric coefficient ε in the absence of salt. The values are calculated using Coulomb's law based on point charges of +0.5 and -0.5. The dashed line represents the baseline thermal energy at 298 K, k_BT, where k_B is the Boltzmann constant and T is the temperature.

3. Theory and modeling of electrostatics in protein engineering

Mechanical modeling of protein structure and dynamics requires a *force field*, a set of atomic and inter-atomic parameters that define how atoms interact in the context of the macromolecule. Among these parameters are the radius and partial charge of specific atom types, optimal bond lengths for atom pairs or optimal angles between groups of three atoms. These parameters are combined in an *objective energy function* which includes bonding (covalent) and nonbonding (electrostatics, van der Waals, hydrogen bonding) interactions to reflect the stability of a specific protein configuration:

$$E = \sum_{bonds} K_r \left(r - r_{eq}\right)^2 + \sum_{angles} K_\theta \left(\theta - \theta_{eq}\right)^2 + \sum_{dihedrals} \frac{V_n}{2}\left[1 + cos(n\varphi - \gamma)\right] + \sum_{atoms\, i<j} \left[\frac{A_{ij}}{r_{ij}^{12}} - \frac{B_{ij}}{r_{ij}^{6}} + \frac{q_i q_j}{\varepsilon r_{ij}} \right] \quad (2)$$

The first three summations incorporate harmonic or periodic potentials for bond vibrations, bond angle constraints and dihedral (bond rotation) constraints. The final term describes nonbonding interactions, including a van der Waals term that prevents atomic clashes, and an electrostatic term. A detailed explanation of the terms and coefficients for this objective energy function can be found in Cornell (Cornell *et al.* 1995).

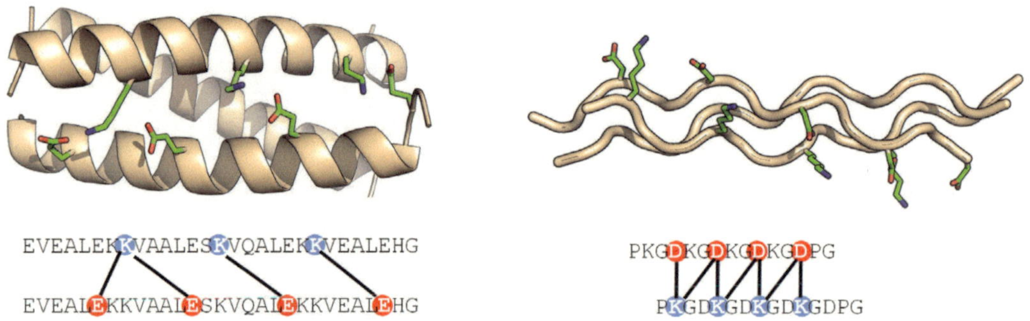

Fig. 2. Charge-pair interactions can be inferred from the sequence for fibrous proteins with periodic structure. (LEFT) The seven-residue heptad of the repeat of the α-helix coiled-coil places acidic (red) and basic (blue) amino acids adjacent in structure. In this case, the interaction between two chains of a three-chain homotrimeric protein are shown (Ogihara *et al.* 1997). (RIGHT) The collagen triple-helix is another type of periodic structure where charge-pairs adjacent in structure can be inferred directly from the sequence. A theoretical model structure of two chains in the triple helix are shown highlighting an extensive charge-pair network. E = glutamic acid, D = aspartic acid, K = lysine.

The objective energy function can be used to predict the lowest energy configuration of a protein chain and to model molecular motions over short (nanosecond to microsecond) periods of time. Electrostatic interactions in the protein are calculated as the sum of all pair-wise atomic interactions. In treating electrostatics this way, two assumptions are made: the partial charge is located at the center of mass of the atom (point charge approximation), and other terms of the multipole expansion beyond ion-ion interactions are disregarded. As physical chemical methods for measuring protein structure, thermodynamic stability and molecular motions improve, the corresponding force fields that model these processes are updated, improving their predictive power. In protein engineering, the modeling task becomes significantly more difficult relative to molecular dynamics, as we are not only concerned with determining the optimal molecular configuration, but also varying the amino acid sequence to modulate properties of the protein. Even for small proteins, the number of possible sequences to consider is immense: for example, a 100-residue protein has 20 possible amino acid choices at each position, resulting in a total space of $20^{100} \approx 10^{130}$ sequences. Combine this with the configurational degrees of freedom of the protein chain and it is clear that an enumeration of all possible states is computationally impossible. To circumvent this search problem, a number of simplifications or coarse graining approaches are used, and no single level of chemical accuracy is universally applied to all protein engineering problems. The requirements of the design problem dictate the level of theory to use. We present several models of electrostatics with varying levels of chemical accuracy that are employed in protein engineering.

The simplest electrostatics treatments do not incorporate atomic detail and assign discrete values to classes of interactions. This reduces a three-dimensional modeling problem to one dimension and is most useful in the design of molecules where positions that are adjacent in structure can be directly inferred from the amino acid sequence. This scenario is found in fibrous proteins such as α-helical bundles and collagen fibrils (Spek *et al.* 1998). Due to the structural periodicity of the α-helix and the collagen triple-helix, it is possible to anticipate which sequence positions are adjacent in structure (Figure 2). Using this information, a scoring function can be used to optimize these interactions. In many cases, the interactions are designed manually without computation(Berger *et al.* 1996; Lombardi *et al.* 1996; Bryson *et al.* 1998; Olson *et al.* 2001; Shi *et al.* 2001). Amino acids of opposite charge are introduced at adjacent positions such that the maximal number of charge pairs is satisfied. When the design goal is of sufficient complexity that computational intervention is required discrete scores are assigned to interactions (Nautiyal *et al.* 1995). One simple scoring function recently applied to both collagen and α-helical proteins is:

$$\begin{array}{ll} \mathrm{Arg} / \mathrm{Arg} & +2 \\ \mathrm{Glu} / \mathrm{Glu} & +3 \\ \mathrm{Arg} / \mathrm{Glu} & -1 \end{array} \tag{3}$$

In this scenario, any structurally adjacent arginine (Arg) pairs are penalized by two kcals/mole. The penalty for adjacent glutamates (Glu) is greater in anticipation of their shorter side chains which bring repulsive charges in closer proximity. Only favorable Arg/Glu interactions are rewarded. The total energy for a given sequence is the sum of all residue-pair scores. If the number of sequences to sample is small, sequence-space can be fully searched. For larger design problems, Monte Carlo methods such as simulated evolution are often used (Hellinga and Richards 1994). Because they ignore molecular details, these scores are far from accurate, but they allow the rapid evaluation of large ensembles of sequences. The discrete scoring function in equation 3 has been used to design stable helical oligomers with specific composition – e.g. combining α-helical chains A and B yielded an A_2B_2 tetramer, without forming A_4 or B_4 species (Summa *et al.* 2002). The same scoring function was recently extended to design of collagen heterotrimers where three peptides A, B and C combine specifically to form an ABC heterotrimer (Xu *et al.* 2011). Molecules such as these are now finding applications as synthetic biomaterials where the electrostatic control of self-assembly is responsible for directing the formation of protein fibers (Pandya *et al.* 2000; O'Leary *et al.* 2011).

When it is necessary to include some level of atomic detail in modeling ion-ion interactions, the simplest potential is Coulomb's law:

$$E = 332 \cdot \frac{\bar{q}_i \bar{q}_j}{\varepsilon \cdot r_{ij}} \tag{4}$$

where the interactions of atoms i and j are a function of charge q and the distance of separation r. The constant of 332 converts the units of energy to kcals/mole. This can be applied to all atoms in the protein as described in equation 2, or restricted to side chains with a net formal charge. In full-atom implementations, the point charge is located at the center-of-mass of the atom, whereas residue-level charges are often placed at the center of the chemical moiety carrying the partial charge, i.e. the center of the guanidino group of the arginine sidechain. The choice of charge is determined by the force field used.

The strength of a charge-charge interaction is influenced by the polarity of the surrounding medium which is reflected by the choice of dielectric coefficient used. In cases where the structural context is known, often a fixed constant dielectric (e.g. 5-10 for the protein interior and ~78 for the surface) is used. One empirical approximation is to use a distance-dependent dielectric ($\varepsilon = 40\, r_{ij}$) based on the premise that the greater the separation between atoms, the more solvent can access the intervening space and screen electrostatic forces (Mayo *et al.* 1990; Gordon *et al.* 1999). In cases where it is desirable to include the effect of counterions, Debye-Huckel and Coulombic terms can be combined to include an ionic strength parameter (Lee *et al.* 2002).

In addition to charge-charge interactions within the protein, solvent-protein interactions are an important electrostatic component of the free energy of folding. Burial of charged side chains in the protein core comes at the cost of desolvating the sidechain ion. These energies can be modeled with reasonable accuracy using finite difference methods applied to the Poisson-Boltzmann equation (Sharp and Honig 1990), but are infrequently used in protein design applications due to the computational burden. Many software packages dedicated to protein design use an atom or residue-level solvation energy that scales with the fraction of accessible surface area buried upon folding. Although these are grossly approximate calculations, rapid algorithms for calculating solvent-exposed surface area make them attractive for evaluating large numbers of candidate sequences. A number of analytic and empirical methods continue to be developed that are finding applications in protein modeling and design (Flohil *et al.* 2002; Morozov *et al.* 2003; Pokala and Handel 2004; Jaramillo and Wodak 2005; am Busch *et al.* 2008). The assumption that atoms have point charges localized to the atom center of mass becomes problematic when designing proteins where electronic polarizability is important, such as the design of metalloproteins where the solvent reorganization energy around the metal can be important for tuning redox properties (Papoian *et al.* 2003), and enzymes where accurate modeling of the transition state and surrounding ligands is critical for an effective design (Tantillo *et al.* 1998). In this case, the use of quantum mechanics calculations is warranted. This approach has been used in the engineering of novel protein catalysts where the active site and substrate transition state are modeled using semi-empirical density functional methods (DFT), and the remaining protein treated using standard molecular mechanics and knowledge based potentials (Jiang *et al.* 2008; Rothlisberger *et al.* 2008).

Modeling hydrogen bonding with reasonable accuracy is an important challenge in protein design. Although primarily electrostatic in nature, hydrogen bonds also have partial covalent character which mediates their linearity in molecular structures. Such properties can only be modeled using quantum mechanical (QM) methods, which is computationally infeasible as these are distributed throughout the protein. Instead, empirical functions are often used that include both proximity and orientation terms. These have been refined using the extensive database of high-resolution protein structures to develop knowledge-based potentials that can capture subtle properties (Kortemme *et al.* 2003). QM methods can also be used to explore the role of other types of electrostatic interactions such as cation-π interactions between ions and aromatic amino acids. In the next sections, several examples of protein engineering of electrostatic properties are presented, highlighting the application of various levels of theory as needed to achieve the design objective.

4. Surface charges in protein electrostatics

It was long thought that surface electrostatics do not make a significant contribution to protein stability because the interactions of polar residues with water in the unfolded state are as energetically favorable as their interactions with each other in the folded state. However, recent work has demonstrated that surface charge optimization can offer significant stability increases to a wide range of proteins (Schweiker and Makhatadze 2009). Surface charge optimization is an attractive option for protein engineering and design because surface positions are generally much more permissive to mutation compared to buried positions, where side chains are prone to clashing as they pack tightly into the protein core. Nature also takes advantage of this evolutionary flexibility at surface positions, modifying surface charge interactions to modulate energetic folding barriers (Halskau *et al.* 2008) and to stabilize thermophilic proteins.

The hypothesis that surface electrostatics can be important for stability is supported by the observation that thermophilic proteins generally contain more charged surface residues than their mesophilic analogs (Kumar and Nussinov 2001). Thermophilic proteins have evolved to be active at high temperatures, and their structures must therefore be very stable. This stabilization is achieved through a number of different strategies, including enriching the sequences in charged surface residues and buried hydrophobic residues at the expense of polar residues. This adaptive response to evolutionary pressure for increased stability has been reproduced in computer simulations of simple lattice model proteins (Berezovsky *et al.* 2007). As a result, the number of salt bridges in a protein is correlated with the temperature of the environment in which its host organism lives (Kumar *et al.* 2000). In one study, mutations to two surface residues of a mesophilic cold shock protein (one of which eliminated an unfavorable electrostatic interaction) yielded a mutant that nearly matched the stability of the thermophilic version of the protein (Perl *et al.* 2000). The stability change was greatly reduced in the presence of 2M NaCl, confirming the importance of electrostatic interactions (which are sensitive to the screening effects of salt) in stabilizing the mutant.

Designed and engineered proteins can also benefit from the stability gains that are possible by optimizing surface electrostatics. The generality of this strategy for protein stabilization was demonstrated experimentally in a study in which the surface residues of a diverse set of five proteins were modified (Strickler *et al.* 2006). A computational algorithm was used to search for mutations to surface positions that would provide the maximum improvement to the energy by adding favorable interactions or alleviating unfavorable ones. Because the combinatorial space of possible surface charges is too large to cover exhaustively, a genetic algorithm was used to search for near-optimal sequences. Genetic algorithms efficiently sample sequence space by mimicking the natural evolutionary process. A population of sequences is generated and evaluated with an energy function - in this case, the energy function was based on a solvent accessibility-corrected Tanford-Kirkwood model. The top-scoring sequences are kept, multiplied, and diversified by random mutations within sequences and crossover or recombination events in which sections are swapped among multiple sequences. At the end of the process, sequences containing between three and eight mutations were selected. One to three designs were constructed for each target protein, synthesized, and purified. Protein unfolding was then monitored by circular dichroism spectroscopy. Remarkably, an increase in stability relative to the wild-type was observed for each of the designed sequences. The largest increase in stability was 4.4 kcal/mol. Another recent study applied this approach to the surface electrostatics optimization of two enzymes. The activity of enzymes is often highly sensitive to even small perturbations to the active site. Nonetheless, human acylphosphatase (AcPh) and human cell-division cycle 42 factor (Cdc42) were successfully stabilized by surface charge optimization with no loss in enzymatic activity (Gribenko *et al.* 2009). Mutant sequences were chosen that maximized the improvement in electrostatic energy while limiting the number of mutations from the wild-type sequence to ~5% of the total residues. The stability of each modified protein was ~10°C higher than their corresponding wild-type protein, while the structures, monomeric nature, and enzymatic activities were retained. This study demonstrated the possibility of increasing the stability of an enzyme by making rational mutations to surface residues on the basis of electrostatic calculations, without disturbing the protein core or the enzymatic activity.

In addition to influencing the intramolecular stability of engineered proteins, electrostatics are important in intermolecular interactions. The balance of charged and hydrophobic residues in a protein sequence is important in determining the tendency of that sequence to aggregate when unfolded (Calamai *et al.* 2003; Chiti *et al.* 2003; Pawar *et al.* 2005). Charged residues within otherwise hydrophobic regions can act as "sequence breakers" that prevent those regions from aggregating. The ability of like-charge repulsive interactions to discourage aggregation is the basis of a surface electrostatics engineering strategy called "supercharging" (Lawrence *et al.* 2007). Amino acids at surface positions of a supercharged protein are mutated to charged residues so that the net charge of the protein is maximized. Net positive and net negative supercharged proteins have both been shown to be less prone to aggregation than their corresponding wild-types. For example, the green fluorescent protein (GFP) is unable to refold into a fluorescent state after thermal denaturation because of aggregation with neighboring unfolded chains. However, the extremely high net charges of supercharged GFP chains disfavor interactions with other unfolded chains of like charge (Figure 3). When a GFP variant supercharged to a net charge of +36 was thermally or chemically denatured, the sample was able to regain up to 62% of its initial fluorescence, confirming that the high net charge of the protein disfavored interchain aggregation.

Streptavidin and glutathione-S-transferase were also successfully supercharged to yield highly aggregation-resistant engineered variants. The supercharging process often involves a relatively large number of mutations, but because the hydrophobic core of the protein is undisturbed, protein folding is typically not significantly adversely affected. For example, the -7 net charge of superfolder GFP was pushed to extremes of +48 (by 36 mutations) or -30 (by 15 mutations). Remarkably, despite the repulsion that would be expected from gathering so many like charges on the surface of a protein, and the stability to be gained by optimizing surface charges demonstrated by the studies presented earlier, the supercharged GFPs were able to fold and fluoresce normally, with only slight decreases in thermodynamic stability. The destabilizing effect of the high concentration of like charges at the surface may be limited by equal or greater destabilization of competing states within the denatured state ensemble (Pace *et al.* 2000). The intuitive electrostatics-based supercharging strategy has already become a popular choice among protein engineers for stabilizing *de novo* designed proteins and therapeutic peptides against aggregation.

Another major limitation of peptide therapeutics is the difficulty of transporting peptides and proteins across the cell membrane. Currently, the leading strategy to improve cellular uptake is to express the target protein as a fusion with one of several polycationic amino acid sequences derived from natural cell-penetrating peptides (Heitz *et al.* 2009). In a recent study, positively supercharged GFP was shown to be capable of entering a range of mammalian cells, and of delivering fused protein payloads more effectively than the standard cationic fusion tags Tat, Arg_{10}, and penetratin (Cronican *et al.* 2010).

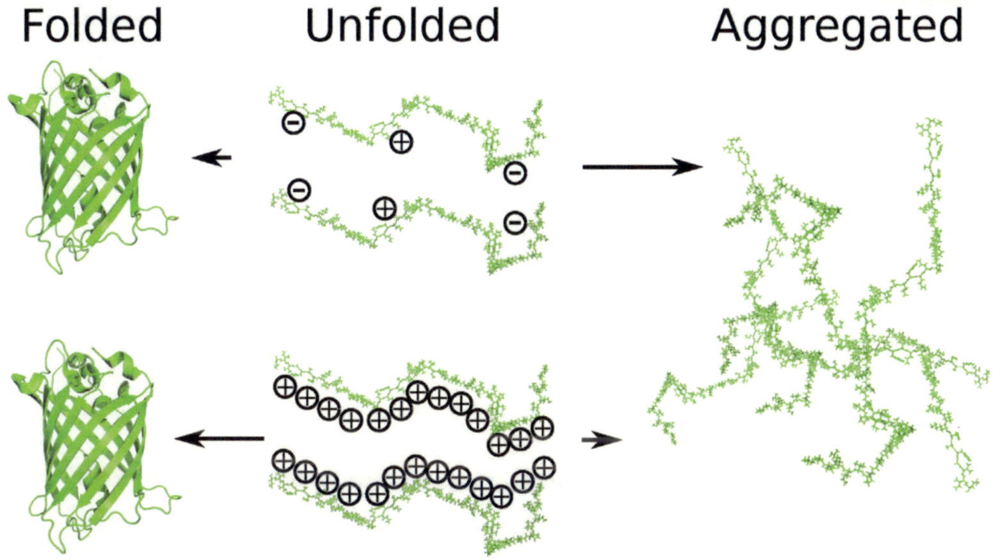

Fig. 3. Supercharging decreases the tendency of unfolded proteins to aggregate by increasing like-charge repulsion. Thermally denatured green fluorescent protein (center) is capable of refolding into the fluorescent state (left) or aggregating with other unfolded chains (right). In the case of the wild-type protein (top), aggregation dominates. In contrast, a sample of a supercharged version of GFP with a net charge of +36 (bottom) regained 62% of its fluorescence following thermal denaturation. The like-charge repulsion between the positive charges on each denatured supercharged polypeptide mitigated aggregation.

5. Electrostatics with buried polar or charged residues

Proteins can tolerate the burial of ionizable residues when environmental modification of the pK_a of the buried side chains prevents them from assuming the charged state. In a series of studies using 96 variants of an engineered form of staphylococcal nuclease, hydrophobic buried residues were individually mutated to lysine, glutamate, or aspartate (Isom *et al.* 2008; Isom *et al.* 2010; Isom *et al.* 2011). The apparent pK_a values of these residues were determined by curve fitting plots of the changes in free energy associated with individually charging the mutants relative to a reference state as a function of pH. In general, the pK_a values of these buried residues were shifted by the environment so that they existed in neutral form within the hydrophobic core.

Protein function can be improved by the burial of a polar residue if the conformation of an associated ligand can be stabilized by electrostatic interactions. Enhanced cyan fluorescent protein (ECFP) was optimized as a FRET[1] donor molecule by mutating several of its residues - S72A[2], Y145A and H148D (Rizzo *et al.* 2004; Malo *et al.* 2007). The new protein variant was called Cerulean. The authors describe the contribution of the H148D substitution of Cerulean in stabilizing a single conformation (i.e., the *cis*-form) of its associated chromophore. Unlike the histidine residue in ECFP, the buried aspartate side group stabilized the *cis*-conformation of the internal chromophore as part of an extended network of hydrogen bonding which included forming a bifurcated hydrogen bond with the indole nitrogen of the chromophore (Figure 4A). The pK_a of the buried aspartate was estimated to be ~6 allowing the residue to remain protonated (i.e., neutral form) for hydrogen bonding, and the smaller size of the aspartate (relative to the histidine) aided in packing of the core. Other hydrogen-bonding interactions were made with nearby polar side groups and with bound water which provided a cage-like enclosure for the internal chromophore (not shown in the figure). The *cis*-conformation of the chromophore placed the six-membered ring of the indole in close proximity to the imidazolinone ring, enhancing energy transfer. The result of the H148D substitution was an engineered molecule that had relatively homogeneous exponential fluorescence emission decay, a property which is necessary for fluorescence-lifetime measurement studies.

The burial of an ionized residue in a protein is an unfavorable event that can be countered by stabilizing electrostatic interactions such as the formation of hydrogen bonding networks. The enzyme ribonuclease T1 is an example of a protein that contains an ionized buried residue, D76, that lacks an ion-pairing partner with which it can form a stabilizing salt bridge. The measured pK_a of D76 is extremely low (pK_a~0.5), ensuring that it always remains fully charged. As a result, it forms a hydrogen bonding network with nearby polar residues T91, Y11, and N9, and with bound water molecules in the protein (Giletto and Pace 1999). This local conformation is depicted in Figure 4B. The wild-type ribonuclease has been shown to have better thermal and chemical stability when compared to uncharged variants D76N, D76S and D76A of the enzyme. In this instance, having a buried charge within a polar microenvironment is advantageous.

---

[1]FRET = Förster Resonance Energy Transfer

[2]The standard one-letter code is used to designate the amino acid residues

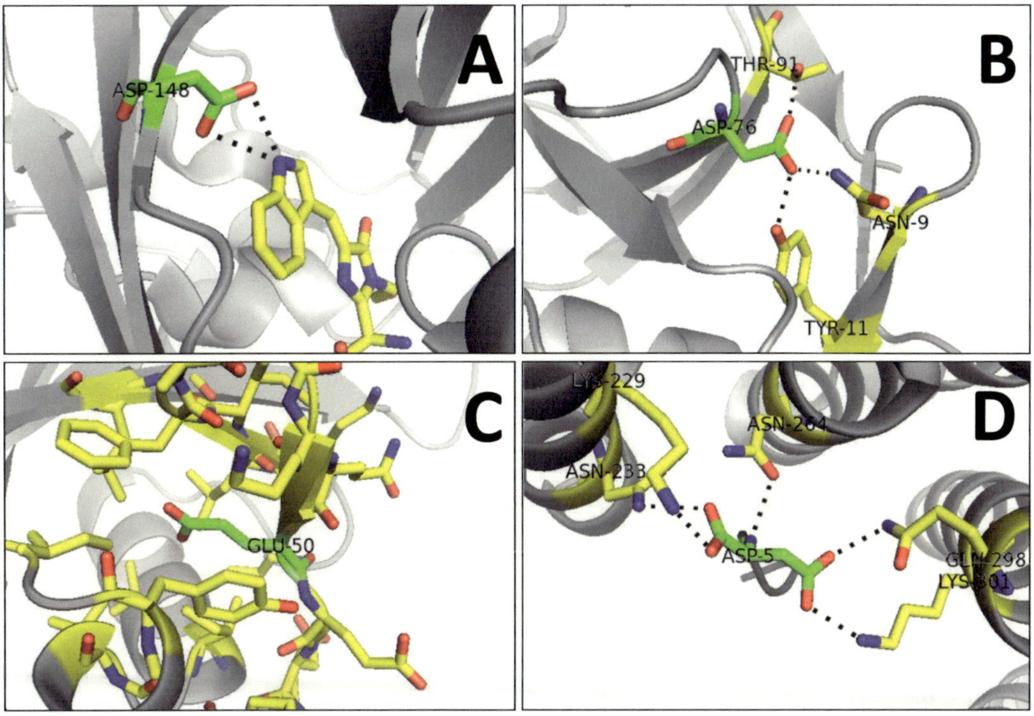

Fig. 4. (A) The key placement of a buried titratable residue can enhance protein function. The cis-conformation of the indole ring of the Cerulean chromophore (shown in yellow) is stabilized by a substituted buried aspartate (shown in green) (PDB code 2q57). The chromophore is comprised of two rings, an indole and an imidazolinone, connected by a methylene bridge. The structure is further stabilized by a network of hydrogen bonding with backbone residues and bound water molecules surrounding the chromophore (not shown). (B) A buried ionized residue that is unable to form a salt-bridge can be stabilized by a hydrogen-bonding network. A charged aspartate residue (shown in green) is stabilized by a network of hydrogen bonding among three polar residues (shown in yellow) within the hydrophobic core of ribonuclease T1 (PDB code 9rnt). (C) The burial of a charged residue can be used to destabilize the protein structure. Glutamate is substituted for leucine (L50E; shown in green) within the hydrophobic core of ubiquitin (PDB code 1ubq). The buried glutamate is surrounded by a hydrophobic microenvironment (shown as yellow residues within 8 Å). The ionization of glutamate results in unfavorable conditions for the charged residue leading to local unfolding in the protein. This charge burial strategy was used to stabilize high-energy folding intermediates of ubiquitin. (D) Residues that become buried following protein-protein interaction can form stabilizing hydrogen-bonding networks. A buried two-carboxylate aspartate of Hsp90 C-terminal peptide (shown in green) is stabilized through its interactions with the polar residues of HOP (shown in yellow) at the protein interface (PDB code 1elr). The Hsp90 peptide is further stabilized along its length by hydrogen bonding with the side chains of the HOP helices (not shown). All figures are generated with PyMOL (Schrodinger, LLC) using a color scheme of red for oxygen and blue for nitrogen, and black dotted lines are used to indicate hydrogen bonding. Hydrogen atoms are not explicitly shown.

Native proteins can be unfolded by ionized residues buried within the hydrophobic core if there are no stabilizing electrostatic interactions to counter the charge. This notion was exploited in a charge burial strategy where "foldons" (regions of secondary structures that cooperatively unfold) of ubiquitin were selectively destabilized in order to trap high-energy intermediate folded states of the protein (Zheng and Sosnick 2010). A strategically located hydrophobic buried residue was substituted with glutamate which was subsequently ionized (Glu^-) during pH titration. In the case of an L50E substitution located at the C-terminal end of the β5 strand, the Glu^- was placed in a microenvironment that was dominantly hydrophobic, with no nearby polar residues or backbone nitrogens to stabilize the charge (Figure 4C). Structural change within the protein caused by the Glu^- was monitored by nuclear magnetic resonance spectroscopy where the authors were able to detect the sequential unfolding of the β5 strand and an adjacent 3_{10}-helix. Interestingly, these unfolded intermediates could be stabilized by pH, and it was possible to refold the protein back to its native structure by neutralizing Glu^-.

Protein-protein interfaces rely on electrostatic interactions to stabilize their previously exposed charged or polar residues. A study on the binding interaction between heat shock protein (Hsp)-organizing protein (HOP) domain TPR2A and the C-terminal end of Hsp90 (MEEVD) revealed the formation of an extensive network of hydrogen bonding between the ionized residues on Hsp90 and polar groups on TPR2A (Kajander *et al.* 2009). As an example, we illustrate the stabilization of the two carboxyl groups of the C-terminal aspartate residue, which is clamped by polar side chains from the TPR2A α-helices and forms hydrogen bonds with K229, N233, Q298 and K301 (Figure 4D). Of these polar groups, N233 was found to be one of several significant binding surface residues that become buried. Similar electrostatic interactions were found along the length of the binding cavity, demonstrating how interfacial residues are stabilized.

6. Acknowledgements

This work was supported by grants from the National Institute of Health (5R21AI088627, 5R01GM089949, 1DP2OD006478 and 1F32GM099291) and the National Science Foundation (DMR0907273).

7. References

am Busch, M. S., A. Lopes, et al. (2008). Testing the Coulomb/Accessible Surface Area solvent model for protein stability, ligand binding, and protein design. BMC Bioinformatics 9, 148.

Anslyn, E. V. and D. A. Dougherty (2006). Modern physical organic chemistry. Sausalito, CA, University Science.

Bashford, D. and M. Karplus (1990). Pkas of Ionizable Groups in Proteins - Atomic Detail from a Continuum Electrostatic Model. Biochemistry 29(44), 10219-10225.

Berezovsky, I. N., K. B. Zeldovich, et al. (2007). Positive and negative design in stability and thermal adaptation of natural proteins. PLoS Comput Biol 3(3), e52.

Berger, J. S., J. A. Ernst, et al. (1996). Stabilization of helical peptides by mixed spaced salt bridges. Journal of Biomolecular Structure & Dynamics 14(3), 285-291.

Bosshard, H. R., D. N. Marti, et al. (2004). Protein stabilization by salt bridges: concepts, experimental approaches and clarification of some misunderstandings. J Mol Recognit 17(1), 1-16.

Bryson, J. W., J. R. Desjarlais, et al. (1998). From coiled coils to small globular proteins: design of a native-like three-helix bundle. Protein Sci 7(6), 1404-1414.

Calamai, M., N. Taddei, et al. (2003). Relative influence of hydrophobicity and net charge in the aggregation of two homologous proteins. Biochemistry 42(51), 15078-15083.

Chiti, F., M. Stefani, et al. (2003). Rationalization of the effects of mutations on peptide and protein aggregation rates. Nature 424(6950), 805-808.

Cornell, W. D., P. Cieplak, et al. (1995). A 2nd Generation Force-Field for the Simulation of Proteins, Nucleic-Acids, and Organic-Molecules. J Am Chem Soc 117(19), 5179-5197.

Cronican, J. J., D. B. Thompson, et al. (2010). Potent delivery of functional proteins into Mammalian cells in vitro and in vivo using a supercharged protein. ACS Chem Biol 5(8), 747-752.

Flohil, J. A., G. Vriend, et al. (2002). Completion and refinement of 3-D homology models with restricted molecular dynamics: Application to targets 47, 58, and 111 in the CASP modeling competition and posterior analysis. Proteins-Structure Function and Genetics 48(4), 593-604.

Giletto, A. and C. N. Pace (1999). Buried, charged, non-ion-paired aspartic acid 76 contributes favorably to the conformational stability of ribonuclease T-1. Biochemistry 38(40), 13379-13384.

Gordon, D. B., S. A. Marshall, et al. (1999). Energy functions for protein design. Current Opinion in Structural Biology 9(4), 509-513.

Gribenko, A. V., M. M. Patel, et al. (2009). Rational stabilization of enzymes by computational redesign of surface charge-charge interactions. Proc Natl Acad Sci U S A 106(8), 2601-2606.

Halskau, O., Jr., R. Perez-Jimenez, et al. (2008). Large-scale modulation of thermodynamic protein folding barriers linked to electrostatics. Proc Natl Acad Sci U S A 105(25), 8625-8630.

Heitz, F., M. C. Morris, et al. (2009). Twenty years of cell-penetrating peptides: from molecular mechanisms to therapeutics. Br J Pharmacol 157(2), 195-206.

Hellinga, H. W. and F. M. Richards (1994). Optimal Sequence Selection in Proteins of Known Structure by Simulated Evolution. Proceedings of the National Academy of Sciences of the United States of America 91(13), 5803-5807.

Isom, D. G., B. R. Cannon, et al. (2008). High tolerance for ionizable residues in the hydrophobic interior of proteins. Proceedings of the National Academy of Sciences of the United States of America 105(46), 17784-17788.

Isom, D. G., C. A. Castaneda, et al. (2011). Large shifts in pK(a) values of lysine residues buried inside a protein. Proceedings of the National Academy of Sciences of the United States of America 108(13), 5260-5265.

Isom, D. G., C. A. Castaneda, et al. (2010). Charges in the hydrophobic interior of proteins. Proceedings of the National Academy of Sciences of the United States of America 107(37), 16096-16100.

Jaramillo, A. and S. J. Wodak (2005). Computational protein design is a challenge for implicit solvation models. Biophys J 88(1), 156-171.

Jelesarov, I. and A. Karshikoff (2009). Defining the role of salt bridges in protein stability. Methods Mol Biol 490, 227-260.

Jiang, L., E. A. Althoff, et al. (2008). De novo computational design of retro-aldol enzymes. Science 319(5868), 1387-1391.

Kajander, T., J. N. Sachs, et al. (2009). Electrostatic Interactions of Hsp-organizing Protein Tetratricopeptide Domains with Hsp70 and Hsp90 COMPUTATIONAL ANALYSIS AND PROTEIN ENGINEERING. Journal of Biological Chemistry 284(37), 25364-25374.

Kaushik, J. K., S. Iimura, et al. (2006). Completely buried, non-ion-paired glutamic acid contributes favorably to the conformational stability of pyrrolidone carboxyl peptidases from hyperthermophiles. Biochemistry 45(23), 7100-7112.

Kortemme, T., A. V. Morozov, et al. (2003). An orientation-dependent hydrogen bonding potential improves prediction of specificity and structure for proteins and protein-protein complexes. Journal of Molecular Biology 326(4), 1239-1259.

Kukic, P. and J. E. Nielsen (2010). Electrostatics in proteins and protein-ligand complexes. Future Med Chem 2(4), 647-666.

Kumar, S. and R. Nussinov (2001). How do thermophilic proteins deal with heat? Cell Mol Life Sci 58(9), 1216-1233.

Kumar, S. and R. Nussinov (2002). Close-range electrostatic interactions in proteins. Chembiochem 3(7), 604-617.

Kumar, S., C. J. Tsai, et al. (2000). Factors enhancing protein thermostability. Protein Eng 13(3), 179-191.

Lawrence, M. S., K. J. Phillips, et al. (2007). Supercharging proteins can impart unusual resilience. J Am Chem Soc 129(33), 10110-10112.

Lee, K. K., C. A. Fitch, et al. (2002). Distance dependence and salt sensitivity of pairwise, coulombic interactions in a protein. Protein Sci 11(5), 1004-1016.

Lombardi, A., J. W. Bryson, et al. (1996). De novo design of heterotrimeric coiled coils. Biopolymers 40(5), 495-504.

Malo, G. D., L. J. Pouwels, et al. (2007). X-ray structure of cerulean GFP: A tryptophan-based chromophore useful for fluorescence lifetime imaging. Biochemistry 46(35), 9865-9873.

Mayo, S. L., B. D. Olafson, et al. (1990). Dreiding - a Generic Force-Field for Molecular Simulations. Journal of Physical Chemistry 94(26), 8897-8909.

Morozov, A. V., T. Kortemme, et al. (2003). Evaluation of models of electrostatic interactions in proteins. Journal of Physical Chemistry B 107(9), 2075-2090.

Nautiyal, S., D. N. Woolfson, et al. (1995). A designed heterotrimeric coiled coil. Biochemistry 34(37), 11645-11651.

Neves-Petersen, M. T. and S. B. Petersen (2003). Protein electrostatics: a review of the equations and methods used to model electrostatic equations in biomolecules--applications in biotechnology. Biotechnol Annu Rev 9, 315-395.

O'Leary, L. E., J. A. Fallas, et al. (2011). Multi-hierarchical self-assembly of a collagen mimetic peptide from triple helix to nanofibre and hydrogel. Nat Chem 3(10), 821-828.

Ogihara, N. L., M. S. Weiss, et al. (1997). The crystal structure of the designed trimeric coiled coil coil-VaLd: implications for engineering crystals and supramolecular assemblies. Protein Sci 6(1), 80-88.

Olson, C. A., E. J. Spek, et al. (2001). Cooperative helix stabilization by complex Arg-Glu salt bridges. Proteins-Structure Function and Genetics 44(2), 123-132.

Pace, C. N., R. W. Alston, et al. (2000). Charge-charge interactions influence the denatured state ensemble and contribute to protein stability. Protein Sci 9(7), 1395-1398.

Pace, C. N., G. R. Grimsley, et al. (2009). Protein ionizable groups: pK values and their contribution to protein stability and solubility. Journal of Biological Chemistry 284(20), 13285-13289.

Pandya, M. J., G. M. Spooner, et al. (2000). Sticky-end assembly of a designed peptide fiber provides insight into protein fibrillogenesis. Biochemistry 39(30), 8728-8734.

Papoian, G. A., W. F. DeGrado, et al. (2003). Probing the configurational space of a metalloprotein core: an ab initio molecular dynamics study of Duo Ferro 1 binuclear Zn cofactor. J Am Chem Soc 125(2), 560-569.

Pawar, A. P., K. F. Dubay, et al. (2005). Prediction of "aggregation-prone" and "aggregation-susceptible" regions in proteins associated with neurodegenerative diseases. Journal of Molecular Biology 350(2), 379-392.

Perl, D., U. Mueller, et al. (2000). Two exposed amino acid residues confer thermostability on a cold shock protein. Nat Struct Biol 7(5), 380-383.

Pokala, N. and T. M. Handel (2004). Energy functions for protein design I: efficient and accurate continuum electrostatics and solvation. Protein Sci 13(4), 925-936.

Rizzo, M. A., G. H. Springer, et al. (2004). An improved cyan fluorescent protein variant useful for FRET. Nature Biotechnology 22(4), 445-449.

Rothlisberger, D., O. Khersonsky, et al. (2008). Novel Kemp Elimination Catalysts by Computational Enzyme Design. Nature.

Schweiker, K. L. and G. I. Makhatadze (2009). A computational approach for the rational design of stable proteins and enzymes: optimization of surface charge-charge interactions. Methods Enzymol 454, 175-211.

Sharp, K. A. and B. Honig (1990). Calculating Total Electrostatic Energies with the Nonlinear Poisson-Boltzmann Equation. J. Phys. Chem. 94, 7684-7692.

Shi, Z., C. A. Olson, et al. (2001). Stabilization of alpha-helix structure by polar side-chain interactions: complex salt bridges, cation-pi interactions, and C-H em leader O H-bonds. Biopolymers 60(5), 366-380.

Spek, E. J., A. H. Bui, et al. (1998). Surface salt bridges stabilize the GCN4 leucine zipper. Protein Science 7(11), 2431-2437.

Strickler, S. S., A. V. Gribenko, et al. (2006). Protein stability and surface electrostatics: a charged relationship. Biochemistry 45(9), 2761-2766.

Summa, C. M., M. M. Rosenblatt, et al. (2002). Computational de novo design, and characterization of an A(2)B(2) diiron protein. J Mol Biol 321(5), 923-938.

Tantillo, D. J., J. Chen, et al. (1998). Theozymes and compuzymes: theoretical models for biological catalysis. Current Opinion in Chemical Biology 2, 743-750.

Xu, F., S. Zahid, et al. (2011). Computational design of a collagen a:B:C-type heterotrimer. J Am Chem Soc 133(39), 15260-15263.

Zheng, Z. and T. R. Sosnick (2010). Protein vivisection reveals elusive intermediates in folding. Journal of Molecular Biology 397(3), 777-788.

Part 3

Measurement and Instrumentation

3

Air-Solids Flow Measurement Using Electrostatic Techniques

Jianyong Zhang
Teesside University
UK

1.Introduction

1.1 Electrostatic charging and discharging

Many industrial processes such as coal pulverising, flour making, cement production, and fertiliser processing involve moving bulk solids by means of pneumatic conveying. Almost all particles become electrically charged during pneumatic transportation, which can be hazardous in industrial environment. The primary sources of electrification are frictional contact charging between particles, between particles and the conducting facility, charge transfer or sharing from one particle to another and charge induction.

Contact charging occurs at their common boundary when two dissimilar substances are brought into contact. On separation, each surface will carry an equal amount of charge with opposite polarity. Triboelectrification can be regarded as a complicated form of contact electrification in which there is transverse motion when two substances impinge or are rubbed together [1]. The transverse motion can in turn accentuate the charge transfer. Contact electrification occurs not only in pneumatic conveying, but also in milling, grinding, sieving and screw feeding.

Another source of electrostatic charge is induction. Charges will be induced on a conductor in an electrostatic field generated by charged particles. This conductor in turn changes the field distribution. If a conductor is insulated from the earth, its potential depends on the amount of charges, the permittivity of particles and their locations relative to the conductor [2]. The charge due to induction disappears when the charged particle moves away from the vicinity or sensing volume of the conductor as in pneumatic conveyance.

Charges can be shared by two particles when they collide to each other, or when one particle is settled on another. Charge sharing is more obvious between conductive particles.

Electrostatic charge can be recombined, for example via the earth or by contact with an object holding opposite charge. However charge on non-conductive particles can be retained and the relaxation time depends on the volume resistivity of bulk solids. If the volume resistivity is high, the charge could be retained even if the solids are in an earthed container. For particles suspended in pure gases as in pneumatic conveying, particles can remain charged for a long period of time irrespective of the particle material's conductivity. Table 1 [3] shows the level of charge accumulation in particles, where the charge carried by unit mass of particle is given for solids of medium volume resistivity emerging from different processes.

Operation	Mass charge density (μC/kg)
Sieving	10^{-5}-10^{-3}
Pouring	10^{-3}-10^{-1}
Scroll feed transfer	10^{-2}-1
Grinding	10^{-1}-1
Micronising	10^{-1}-10^{2}
Pneumatic conveying	10^{-1}-10^{3}
Triboelectrical powder coating	10^{3}-10^{4}

Table 1. Charge build up on powder

Electrostatic charging can be a hazard if the charges are suddenly released via discharging to earth or another body, which produces a high local energy density and thus act as a possible ignition source. In section 7.2.4 of CENELEC and British Standard PD CLC/TR 50404: 2003, the discharges have been classified as "spark discharges", "brush discharges", "corona discharges", "propagating brush discharges", "cone discharges" and "lighting like discharges". Among them, spark discharges, brush discharge and lighting like discharges may occur in pneumatic conveying. The incendivity of discharge is very much depends on the energy stored and the minimum ignition energy (MIE). Therefore the hazardousness of discharge depends on the area classification (zones) and gas group of process environment.

Potential build-up on metal items (pipe lines, flanges, bolts and etc) can be avoided by earthing all these items. Sometimes pipe sections can become floated due to gaskets and other insulators. Therefore it is important to bond such sections to the earthed sections. Care must be taken when non-conductive pipes and hoses have to be used for pneumatic conveying, the maximum possible energy stored must not exceed the MIE. In some case, it is possible to choose the dense phase conveying which can reduce the risk of ignition inside pipe due to lack of air. Different from one's empirical knowledge, humidification is not effective as a means of dissipating the charge from a dust cloud. The precaution for electrostatic discharging is not the main focus in this chapter, more details can be found in British and CENELEC Standard PD CLC/TR 50404:2003, "Electrostatics—code of practice for the avoidance of hazards due to static electricity" [3].

1.2 Brief history of electrostatic techniques for air solid flow measurement

Electrostatic charging of flowing particles has long been known. The method to relate the magnitude of charge to the flow parameters was studied as early as in 1963. Batch, Dalmon and Hignett [4] used a pin probe to detect the current from the probe to earth in order to measure the mass flow rate of pneumatically conveyed pulverised fuel (p.f). The probe current depends on charge generated on probe due to contact electrification and induction.

The derivation of relations between the probe current and flow parameters was based on a model developed by Cooper [5] and Hignett [6] which was for electrification of liquids in motion.

Assume that the current flowing into the probe is I, the general relation is expressed as,

$$I = f_n(\varepsilon, K, d, V, \rho, \dot{M}, A_P) \tag{1.1}$$

Where

ε --the electrical permittivity of particle

K--the electrical conductivity of particle

d--the particle diameter

V--the flow velocity in the region of the probe

ρ--the density of air

$\dot{M}$ --the mass flow rate or flux of particle flow

A_P--the cross-sectional area of probe

According to Pi Theorem [7], these variables may be expressed as four dimensionless groups such that

$$\left(\frac{I^2}{\varepsilon d^2 \rho V^4}\right) = f_n\left[\left(\frac{\varepsilon v}{Kd}\right), \left(\frac{d^2}{A_P}\right), \left(\frac{\dot{M}\varepsilon}{Kd\rho}\right)\right] \tag{1.2}$$

Then if the condition where only V and $\dot{M}$ vary, other parameters are assumed to be constant, the above function can be simplified as

$$\left(\frac{I^2}{V^4}\right) = f_n\left(\dot{M}\right) \tag{1.3}$$

At the time of research, the conclusion was that "the unique relation between probe current and the p.f. flux (mass flow rate) cannot be obtained, thus the electrostatic probe cannot be used as a means of determining the flux of p.f. in a pipe". The method proposed by Batch et al has been known as intrusive method which has been used by Hignett [8], and further explored by Soo [9] and King [10]. A commercial product based on the same principle, which is comprised of three probes installed with 120° gap apart, has been designed and manufactured by TR-tech (now owned by Foster Wheeler) [11] [12].

King [10] has used non-intrusive method which measures the induced voltage on an insulated pipe section (Pipe wall sensor), and he compared AC voltage measurement method to DC measurement method [13] [14]. The AC or noise measurement method, also known as "dynamic" method [15] takes the AC signal component to indicate flow concentration. Coulthard applied this technique to coal flow measurement in the Methil power station in Scotland [16], and Gajewski has employed it to measure dust in motion [17]. Gajewski further studied the measuring mechanism by combining the electrostatic field theory with electrical circuit analysis for a lined circular pipe wall sensor [18]. Since 1970s, positive and convincing results have been reported for measuring a range of different particles, e.g. pf, glass beads, sands and polymer granules, which encouraged further study in this area.

The mechanism of electrostatic metering system has been studied by many people, for example, Gajewski [19], Massen [20] and Hammer [21] studied filtering effect of the circular electrode, and Yan investigated charge induction based on free space electrostatic field theory [15]. When the Finite Element Method became practically viable with the increased speed and capacity of computers, more in depth study of charge induction became possible. A model describing the relation between the induced charge on an insulated pipe section (electrode) and the charge carried by a particle with respect of its location, known as "Spatial sensitivity" was established by Cheng [22]. This model was developed based on electrostatic field theory using the Finite Element method (FEM). Based on the above model, and with further study, Zhang [23] has related flow concentration and solids mass flow rate to the charge level on electrodes by employing stochastic process theory. He also investigated and verified 2-D spatial sensitivity of ring-shaped electrode to "roping" flow [24] and effects of particle size [25] on the charge carried by particles. The effects of velocity and concentration have been studied experimentally and theoretically since then [26]. Cheng's model [22] has also been used by Xu [27], and an exploitation of frequency method for velocity measurement has also been based on the Cheng's model [28] [29]. The product operating according to dynamic electrostatic techniques with trade name PfMaster has been developed and manufactured by ABB Ltd.

2. Charge induction and "dynamic" measurement method

2.1 Mathematic model for charge induction

In this section, the main focus of the analysis will be on the meters with circular electrodes. For the analysis of other types of electrostatic sensors, the same principles apply. The method adopted here is based on the analysis conducted by Cheng [22].

Fig.1 depicts a simplified, schematic view of circular electrostatic meter. The circular metallic electrode is installed flush with, and electrically insulated from the inner surface of the earthed pipe, but exposed to the medium inside the pipe. This arrangement ensures that the electrode is sensitive to the charges carried by particles without restricting flow and can avoid severe charge build-up on non-conductive lining. It also minimises the electrode wear that can occur with intrusive probes [30]. The charge generated on the electrode is due to the following effects:

1. particle contacting with the electrode, and
2. charge induction due to the presence of charged particles within the sensing volume.

Since the speed of charged particle through the sensing volume is insignificant compared to the speed of light, so the electromagnetic effect generated by flowing charged solids can be neglected. The analysis therefore is under the assumption of a pure electrostatic electrical field. The electrode is narrow compared to the pipe diameter, so the contact charge between particle and electrode was not considered. The model presented here is also assumed a lean phase flow regime, under which effect of dielectric property of solids on the electrostatic field is neglected.

The principle of the measurement can be approximately (but not accurately) explained as follows: Regarding the entire conveying pipe (ignoring the insulator) as an enclosed system, the total charge induced on the inner surface of the system should equal to the charge carried by the source particle, but of opposite polarity. The portion of the total induced charge on the electrode varies with the location of the particle (although total charge induced on the inner surface of the enclosed system does not vary).

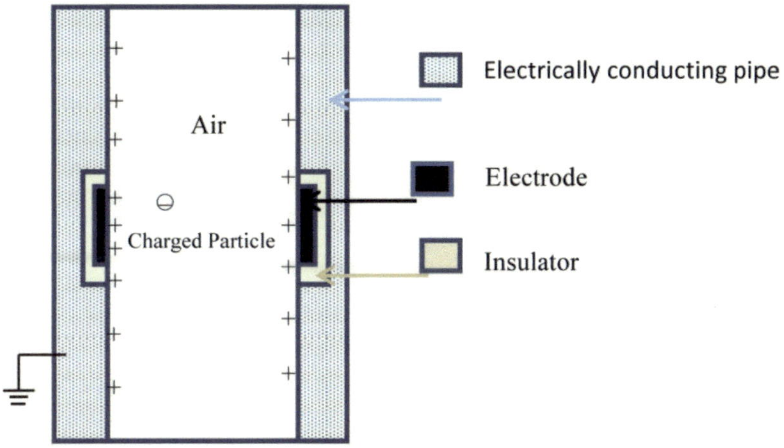

Fig. 1. Charge induction

For the convenience of calculation, assume a non-conductive, negatively charged particle surrounded by air as the only source of electrostatic field.

In order to know the induced charge Q on the electrode, the charge density on the inner surface of the electrode, σ, needs to be found.

$$Q = \int_S \sigma ds \tag{1.4}$$

where S stands for the entire inner surface area of the electrode, s is the surface area variable.

According to the electrostatic theory, the surface charge density is equal to the electric flux density (electric displacement) D, i.e.

$$D = \sigma \tag{1.5}$$

$$\nabla \cdot D = \rho \tag{1.6}$$

$$D = \varepsilon E \tag{1.7}$$

$$E = -\nabla \Phi \tag{1.8}$$

∇ is the gradient operator, E is the electrical field strength, ε, the relative permittivity of the medium and Φ refers to as the electrical potential.

From equations 1.5 to 1.8, Equation 1.9 and 1.10 can be derived,

$$\nabla \cdot D = \nabla \cdot \varepsilon E \tag{1.9}$$

$$\nabla \cdot (\varepsilon \nabla \Phi) = -\rho \tag{1.10}$$

Assume the following boundary conditions

$$\Phi(\Gamma_P) = 0 \bigcup \Phi(\Gamma_i) = 0 \cup \Phi(\Gamma_t) = 0 \tag{1.10}$$

where $\Gamma_P, \Gamma_I, \Gamma_t$ represent the boundaries of the earthed conveying pipe, the insulator and the electrode respectively.

The conveying pipe is earthed, so the potential on it is zero. The electrode is usually connected to a charge amplifier in which the electrode is virtually earthed, the potential on electrode is very close to zero. It may be noticed that the potential on the insulator is hard to set. In the simulation, it was set as zero, and other low voltages (2, and 3Volts) on the insulator the similar results were obtained.

The problem becomes to find the potential Φ. If the potential distribution is known, the charge density on the inner surface of electrode can be found, hence the induced charge on the electrode can be derived from Equation 1.4. This is a 3-D problem. The location of charged particle affects the amount of induced charge on the electrode. However in a cylindrical co-ordinate, if the particle only changes it angular co-ordinate with its radial (r) and axial (x) coordinates keeping unchanged, the induced charge on the electrode should not change due to the symmetrical configuration of the system. Consider also the superposition theorem in electrostatic field, a 2-D model is sufficient for solving this 3-D problem, and a ring-shaped charge situated with its axis coinciding with the pipe central line will produces the same induced charge on the electrode as a point particle carrying the same amount of charge at the same axial and radial locations. The equivalent ring was used by Cheng to calculate the charge induction as shown in Fig.2.

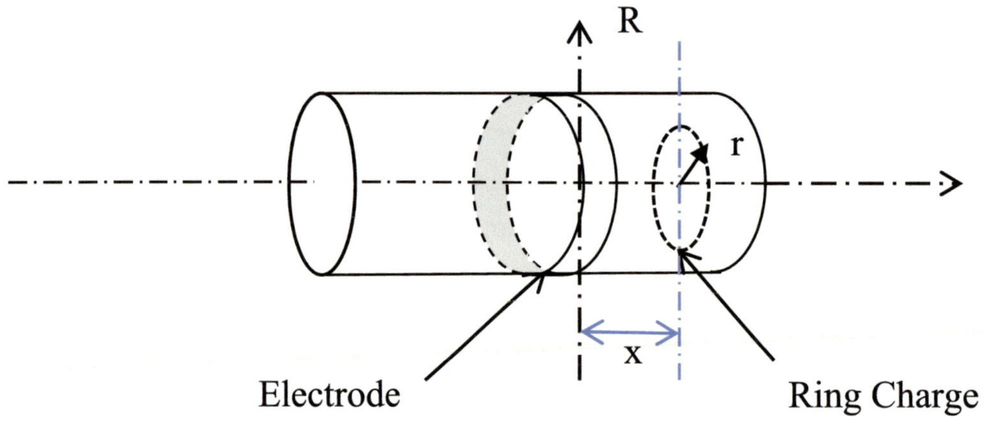

Fig. 2. Charge Induction

The detailed analysis can be found in [22]. Here provided is an equation relating the charge induction and source charge and its location, also known as "spatial sensitivity" which was obtained from FEM simulation and regression.

$$Q = Ae^{-kx^2} \tag{1.11}$$

where Q is the charge induced on the electrode due to a point charged particle carrying unit charge located at (x, r, θ), but Q depends on r and x only. A and k are two parameters determined by electrode geography, namely W/R ratio (where W is the width of electrode, and R the radius of the sensor) and radial location r of the charged particle.

2.2 Spatial sensitivity

Fig. 3 shows the relationship governed by Equation 1.11 for a given electrode W/R ratio (W/R=1/5) when the unit charge particle moves along the pipe axial direction (x coordinate) at different radial locations.

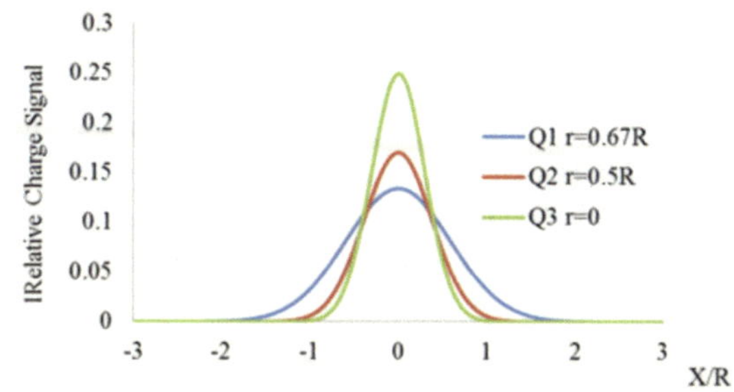

Fig. 3. Spatial Sensitivity for Particle passing along Different Axies.

Fig. 4. compares the spatial sensitivity for the electrodes of different width as a particle move along the pipe central line (r=0)

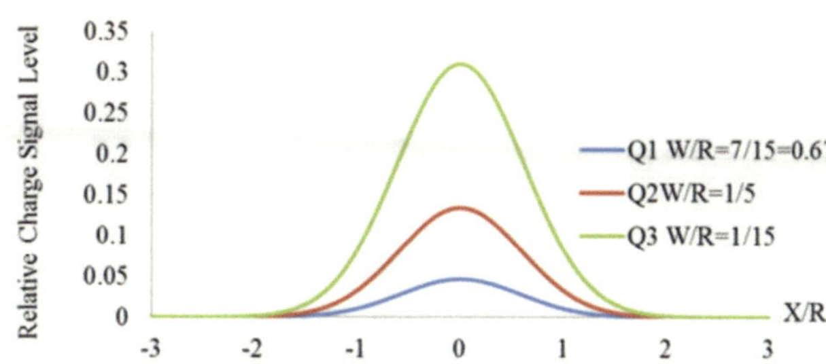

Fig. 4. Spatial Sensitivity for Different Electrode Widhts

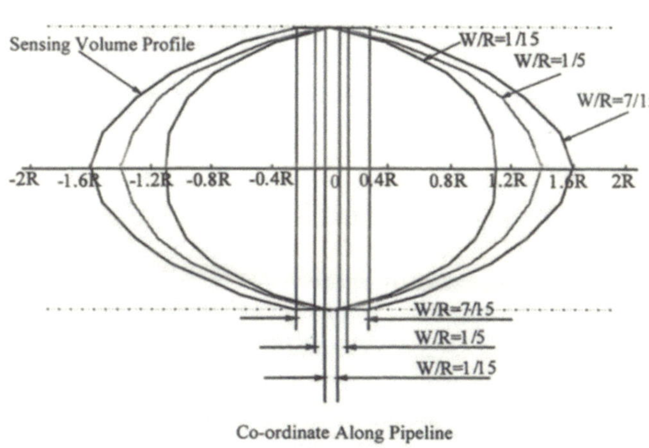

Fig. 5. Sensing Volume of electrodes of different width

Fig. 5. [22] depicts the sensing volume of the electrodes with different width to radius ratios. In the figure, a minimum value of spatial sensitivity has been set. A point is within the sensing zone if the spatial sensitivity at that point is above this value. The shape of sensing volume depends on the geometry of the electrode.

If the particle movement along the axial direction is the main concern, the velocity of the particle in this direction can be related to Equation 1.11 by replacing x with V·t, where V is the particle velocity along the pipe axial direction, t is time.

$$Q = Ae^{-kV^2t^2} \tag{1.12}$$

This temporal expression relates the time, axial velocity and induced charge together, where for a given electrode, A and k vary with radius location r only. The radial velocity component of particle is not considered. The recent research on analysis of radial velocity can be found in [31].

The Fourier transform of Equation 1.12 provides the frequency property of the electrode to a point charge moving at velocity V along the pipe line.

$$Q(\omega) = F\{Q\} = F\left\{Ae^{-kV^2t^2}\right\}$$

Therefore,

$$Q(\omega) = \frac{A}{V}\sqrt{\frac{\pi}{k}}e^{-\frac{1}{4k}\left(\frac{\omega}{V}\right)^2} \tag{1.13}$$

The analysis conducted by Cheng [22] is presented above. Different from the previous analysis, the model in Equation 1.11 has taken the presence of metal conveying pipe (earthed), the insulator between the conveyor and electrode, and the effect of resultant charge on the electrode into account. The significances of this model are that it allows studying the effects of sensor geometry and velocity on charge induction, and permits the frequency analysis. From this model, 2-D and 3-D spatial sensitivity profiles of a sensor can be derived. Equation 1.11 was the first such expression to be used for temporal and frequency domain analysis and it can be used as a guide for sensor design.

2.3 Dynamic measurement

As reviewed in section 1, King [10] compared AC and DC measurement methods for both circular sensor (he named it as "pipe sensor") and pin sensor (intrusive probe). In industrial environment, DC signal on electrodes is more prone to interference so the fluctuation of induced charge has been used for measurement. An electrostatic metering system which measures the signal fluctuation is termed unofficially "dynamic" measurement system, although in the author's view, the word "dynamic" has been misused. In such a system, it is the change or variation of the induced signal that matters. The fluctuation produced by air-solids flow is regarded as band-limited white noise [13] proportional to solids concentration [32]. The fluctuation in number of particles, random movement of particles, particle size and shape changes can also result in the random change in signal level. The signal level is dependent upon mass flow rate or concentration for given mean velocity, the distribution of particle size, humidity and etc.

3. Measurement of velocity and mass flow rate

3.1 Unit impulse response of ring-shaped electrode

Equations 1.11 1.12 and 1.13 provide the temporal and frequency spatial responses of a circular electrostatic meter to a charged particle. Zhang [23] extended these models to study the response to flow concentration and flow mass flow rate. In order to simplify the analysis, it is assumed that the particles of uniform size are evenly distributed over the sensing volume so that the volume concentration of solids is determined only by the number of particles per unit volume, i.e. N. Because the solids are fed or dropped into the conveying system at an upstream point, so that N and solids concentration can be regarded as a waveform travelling along pipe line at velocity V. The point of injection is the source of the wave. Hereafter the number of particles per unit volume and concentration will be respectively denoted as N(x,t) and Con(x,t), both of which depend on x, the axial distance and t, the time. The charge induced on the electrode, Q, is a function of N(x,t) in Equation 1.14 or a function of Con(x, t) as in Equation 1.15.

$$Q = G_n \overline{D}^2 \iint_{Vol} \mathrm{rA(r)N(x,t)e^{-k(r)\varphi^2 x^2}}\, drdx \tag{1.14}$$

$$Q = \frac{G}{\rho_m \overline{D}} \iint_{Vol} \frac{\mathrm{r}}{\mathrm{R}} \mathrm{A(r)Con(x,t)e^{-k(r)\phi^2 x^2}}\, d(\frac{r}{R})dx \tag{1.15}$$

where N(x, t) is the waveform of the number of particles which varies along the pipeline at a given time, and at any point of x, it varies with time; ϕ is a constant for a given diameter of sensor. A and k depend on radius r for an electrode of a given width; G_n and G are constants related to particle surface charge density and pipe geometry. Con(x,t) is the concentration waveform, and R is the radius of the electrode.

In order to find the unit impulse response of the electrode, let Con(x,t) be a delta function, i.e.

$$Con(x,t) = \delta(t - \frac{x}{V}) \tag{1.16}$$

thus there is only one non-zero point at any given time in the co-ordinate x which is V*t (or at given point x, the impulse arrives at time x/V). Under such a concentration, the induced charge is the unit impulse response of the electrode. From Equation1.15, we have

$$Q = h(t) = \frac{G}{\rho_m \overline{D}} \int_0^1 \frac{r}{R} A(r) e^{-k(r)V^2(t*\phi)^2} d\frac{r}{R} \tag{1.17}$$

and the Fourier transfer function of the electrode is

$$H(\omega) = \frac{Q(\omega)}{Con(\omega)} = \frac{G\pi^{\frac{1}{2}}}{\rho_m \overline{D}V} \int_0^1 \frac{r}{R} \frac{A(r)}{\sqrt{k(r)}} e^{-\frac{(\varpi/\varphi)^2}{4V^2}\frac{1}{k(r)}} d\frac{r}{R} \tag{1.18}$$

where Q(ω) is the Fourier transfer function of the induced charge, and the Con(ω) is the Fourier transfer function of the concentration.

3.2 Equivalent circuit and charge amplifier

The signal induced on the electrode has to be connected to a measuring equipment or a preamplifier. Usually the input impedance of a preamplifier or a measurement equipment is finite, therefore the characteristics of the sensor comprising the electrode and the connected electronics depend not only on the electronics but also on the internal impedance of the electrode due to "loading effect".

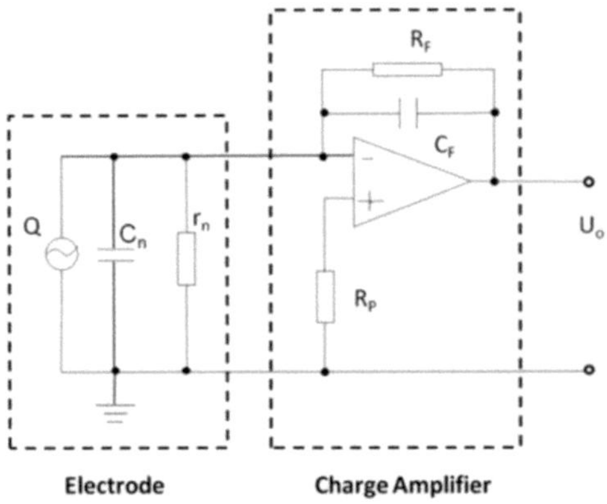

Fig. 6. Sensing system

Although there are various types of preamplifier circuits, the charge preamplifier is among those of most widely used for such systems. As shown in Fig.6 the electrode is at virtual earth potential. The capacitance C_F in the feedback loop is used to suppress the effects of the wiring capacitance and the equivalent capacitance C_n, when the values of the wiring capacitance and C_n as well as their variation have to be considered. R_F is used to provide a DC path, which also determines the lower cut-off frequency with C_F.

The charge amplifier blocks DC component of the signal Q, so the system measures signal fluctuation, performing "dynamic" measurement.

The transfer function of the measurement system T(ω) is

$$T(\omega)=\frac{U_o(\omega)}{Con(\omega)}=\frac{Q(\omega)}{Con(\omega)}\frac{U_o(\omega)}{Q(\omega)}=H(\omega)P(\omega) \tag{1.19}$$

where $U_o(\omega)$ is the output voltage of the charge amplifier, P(ω) is the transfer function of the charge amplifier. The loading effect of source is reflected in P(ω) which depends on C_n and r_n.

3.3 Measuring solids mass flow rate

Assume the concentration signal is a band-limited white noise, based on equation 1.19, Zhang [23] has used Parseval's formula to relate the rms of U_o to the fluctuation in concentration,

$$U_{rms}^2 \propto \frac{con_{rms}^2}{\varpi_{\max}} V \tag{1.20}$$

If the root mean square con_{rms} of the flow noise con(t) is directly proportional to the mean solids concentration $\overline{Con(t)}$, U_{rms}, the rms value of the sensor's output has a linear relationship with the mean solids concentration for given solids density and particle size.

The above analysis assumes that the net charge carried by solids does not depend upon velocity. Hence the velocity in Equation 1.20 reflects the effect of velocity on characteristics of the sensing system only. The amount of net charge carried by particles is actually affected by velocity. Gajewski [17] has studied the this effect on the 'charging tendency' of PVC dust, and Masuda conducted the tests on several different materials and found that the electrostatic meter's output was proportional to V^{γ} [14], where γ varied from 1.4-1.9. The effect of velocity on the net charge has also been confirmed from many tests on pulverized coal [33].

If we assume the net charge is proportional to the solids velocity over the range investigated, as it was suggested by Gajewski [17] for PVC dust over the velocity range below 20m/s, equation 1.20 becomes

$$U_{rms}^2 \propto \frac{con_{rms}^2}{\varpi_{\max}} V^3 \text{ or } U_{rms} \propto \frac{con_{rms}}{\sqrt{\varpi_{\max}}} V^{3/2} = \dot{M}\sqrt{\frac{V}{\omega_{\max}}} \qquad (1.21)$$

where ω_{max} is the signal frequency up limit, $\dot{M}$ is the solids mass flow rate. Hence the root mean square value of signal can be used to directly measure solids mass flow rate if the effect of velocity has been compensated.

3.4 Velocity measurement

In Equations 1.20 and 1.21, it can be seen that, to achieve accurate concentration or flow rate measurement result, the effect of velocity needs to be compensated for, therefore, the velocity measurement cannot be avoided.

There are several different ways to measure velocity of conveyed solids, however, the cross correlation method remains the most practical and viable one. Since late 1960s and early 1970s, the cross-correlation found its applications in flow measurement. Various sensors have been used to measure different types of flow. A cross correlatior detects the flow noise transit time, from which the mean velocity of flow can be derived. Beck [34], Coulthard [35], Cole [13] and King [10] used this method to measure velocity of multi-phase flow. Keech and Coulthard realised a microprocessor based electrostatic cross correlator for the ABB cable meter [36]. Cheng adopted "polarity cross correlation" to measure pulverised coal flow velocity in a blast furnace [37]. The technique has been further improved to accommodate multi-channel velocity measurement [38].

In electrostatic air-solids flow measurement system, usually two identical electrodes are mounted up and down stream with a known distance apart. If the flow concentration Con(t) is rectilinearly transferred from upstream to downstream at a velocity V, it can be expected that the signal from the downstream electrode is a delayed replica of the signal from the upstream electrode, i.e.

$$\mathrm{Con}_2(t) = \mathrm{Con}_1(t - L/V) = \mathrm{Con}_1(t - t), \qquad (1.22)$$

where L is the distance between two electrode. τ is the transit time.

Because L is known, thus once τ is found, the velocity V can be determined from Equation 1.23

$$V = L / t. \tag{1.23}$$

According to the definition, the cross-correlation function between $Con_1(t)$ and $Con_2(t)$ is equal to

$$R_{c_1c_2}(\tau) = \lim_{T\to\infty} \frac{1}{T}\int_0^T \mathrm{Con}_1(t-\tau)\mathrm{Con}_2(t)dt \tag{1.24}$$

If the two signals are exactly identical,

$$R_{c1c2}(\tau) = \lim_{T\to\infty} \frac{1}{T}\int_0^T \mathrm{Con}_1(t-\tau)\mathrm{Con}_1(t-L/V)dt \tag{1.25}$$

The cross correlation becomes a delayed version of auto correlation of $Con_1(t)$, as shown in Fig. 7.

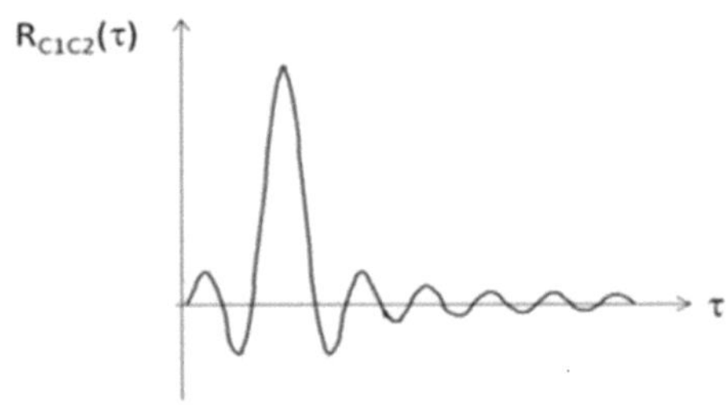

Fig. 7. Cross Correlation

$$R_{c_1c_2}(\tau) = R_{c_1c_1}(\tau - L/V) = R_{cc}(\tau - L/V) \tag{1.26}$$

In reality, two signals are not exactly identical, however the cross correlation efficient can be very high. Even for low cross correlation coefficient, say, 0.5, a cross correlator can still successfully capture the flow transit time and find the average flow velocity. The frequency band of the signal determines the measurement accuracy of transit time, which in turn affects the accuracy of velocity measurement.

4. Relative measurement

The response of an electrostatic meter for air solids flow measurement depends on density, particle size, velocity, mass flow rate and flow profile. Over the past ten years, the performance of dynamic electrostatic meters has been significantly improved, however the high measurement accuracy is still not achievable if all the above parameters vary over wide ranges.

In many cases, only two or three parameters vary and other parameters stay relatively stable. This is particularly true in coal-fired power station, where pulverised fuel comes from a mill and split into six or eight conveyors. Under normal conditions, the density of solids, moisture content and even flow profile are similar in different conveyors, but particle size distribution, mass flow rate and velocity vary from one conveyor to another. If the system can provide the signal proportional to the split (relative or percentage of overall mass flow rate) with velocity and particle size compensation, the mass flow rate in each conveyor can be given with reasonable accuracy because the overall loading entering the mill is known.

4.1 Signal, concentration, mass flow rate and velocity

Fig. 8 presents a typical set of test results on a dynamic electrostatic meter. The tests were carried using the Teesside University 40mm diameter rig, and the material used was "Fillite", a commercial product made from fly ash. The air and solids mass flow rate were controlled to maintain the constant air to solids ratio (i.e., mass flow rate of air/mass flow rate of solids), hence under each of the ratios 3.86, 3.34, 2.88, 2.39 and 1.92, when the solids mass flow rate increases, the air flow rate is increased proportionally. For each air to solids ratio, the relationship between signal rms value and the solids mass flow rate was close to a second order polynomial due to combined effect of solids mass flow rate and velocity [26].

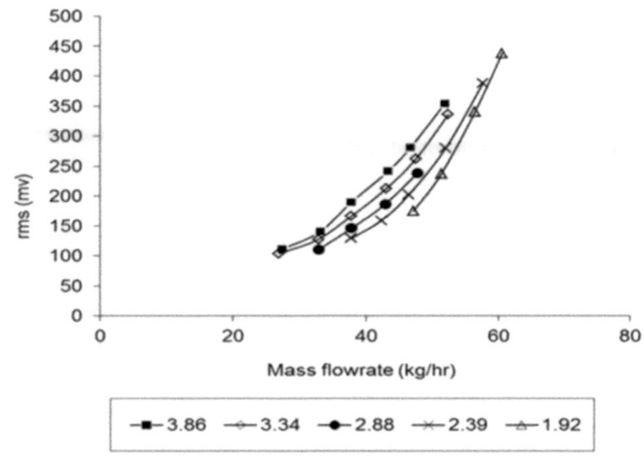

Fig. 8. Response of Electrostatic meter

It is also clear that the higher air to solids ratio (means less solids, or lower concentration) resulted in higher signal for a given mass flow rate. It seems contradictory to the common sense, but again it is due to higher velocity, the hidden information in the graph. The signal is more sensitive to velocity than to any other parameters, and the effect of velocity requires compensation for mass flow rate or concentration measurement.

The signal depends on the combined effects of concentration, mass flow rate and velocity. From the above analysis, an algorithm given by Equation 1.27 was derived to relate the meter's output signal, solid mass flow rate and air to solids ratio (or concentration),

$$U_{orms} = (AR_{as} + B)\dot{M}^2 + (CR_{as} + D)\dot{M} + ER_{as} + F \tag{1.27}$$

where U_{orms} is the rms value of output voltage of the meter, A, B, C, D, E and F are constants, R_{as} represents air to solids ratio, and $\dot{M}$ is the solids mass flow rate. Fig. 9 shows the measured mass flow rate against the true mass flow rate for various velocities and air to solids ratios [26].

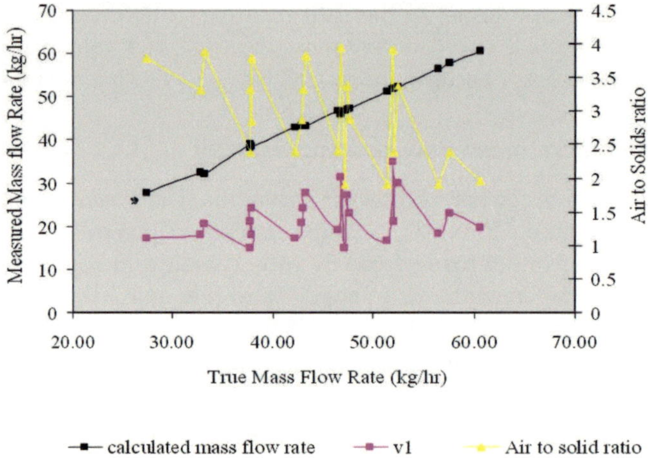

Fig. 9. Calibration Graph

4.2 Effect of particle size

As discussed from the beginning of this chapter, the induced charge on the insulated electrode is a function of several factors including particle size.

From Equations 1.15, 1.17, 1.18, it can be seen that induced charge on electrode is inversely proportional to particle size. It is due to the fact that the mass of solids is proportional to D^3 and the total surface area of solids is proportional to D^2 for spherical particles, where D is particle diameter. If surface charge density is a constant, larger surface of total particles will provide higher signal level when the particles are getting smaller [25].

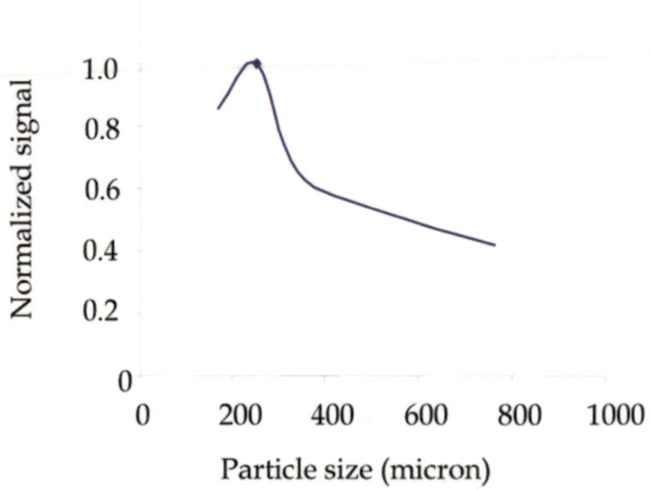

Fig. 10. Signal Vs particle size

Fig. 10 was obtained from experiments using sieved materials [39]. For the given mass flow rate, velocity and concentration, the signal from a dynamic electrostatic meter decreased with particle size for the size above 250 μm, confirming Equation 1.15, 1.17 and 1.18. However for particles below that size, the signal reversed the trend, i.e. the smaller particle size resulted in lower signal. At the time of experiment, the signal drop for smaller particles was thought to be caused by the sudden change in solids flow rate. The recent research revealed that the signal drop for small particles could have been caused by flow regime change. When the size of particle is getting smaller, the flow becomes less turbulent. This effect outweighs the effect of total particle surface area increase so that overall signal level decreases.

4.3 Spatial sensitivity

In pneumatically conveyed air solid flow, the distribution of solid phase is often un-even. For example around bends and restrictive devices, the roping flow regime may be formed. The air solids flow profiles depend on conveying velocity, particle size, humidity and geometry of conveyor. The research in this area can be found elsewhere [40][41][42].

The measurement results will be affected by flow regime unless a meter has a uniform spatial sensitivity.

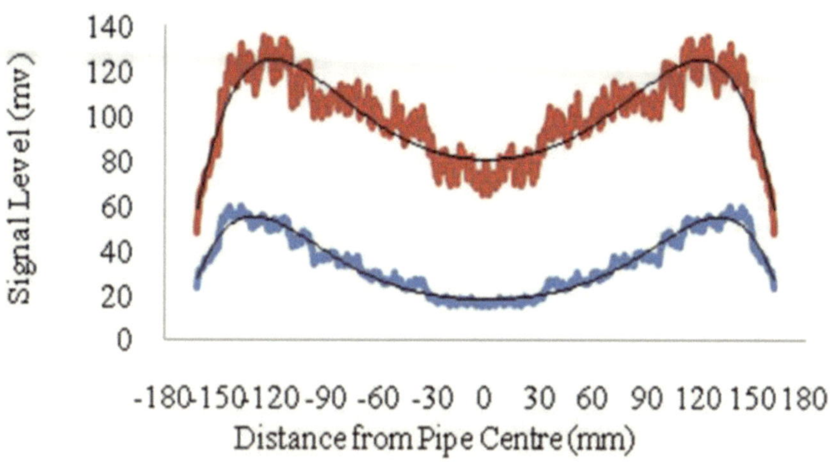

Fig. 11. Spatial Sensitivity Test Results

Fig.11 depicts the test results obtained on a 14" (356mm) diameter electrostatic meter [24]. A roping stream of constant flow rate was provided with an one-inch jet, the roping stream was parallel to the pipe axial central line and moved cross the pipe cross sectional area along its diameter. The material used was pulverised coal. The output voltage (rms) of the meter to the "roping" flow stream was recorded when the jet moved from the centre to a location very close to pipe wall. The signals on the wide electrode (Red W/R=0.5) and on the narrow electrode (Blue W/R=0.014) followed the same trend. It is clear that the signal increased with the flow stream getting closer to the pipe wall, and then it started to drop as the roping stream crossing about 70% of full radius, which is caused by combination of the increased sensitivity and the reduced sensitive volume of the sensor as shown in Figs 3, 4 and 5.

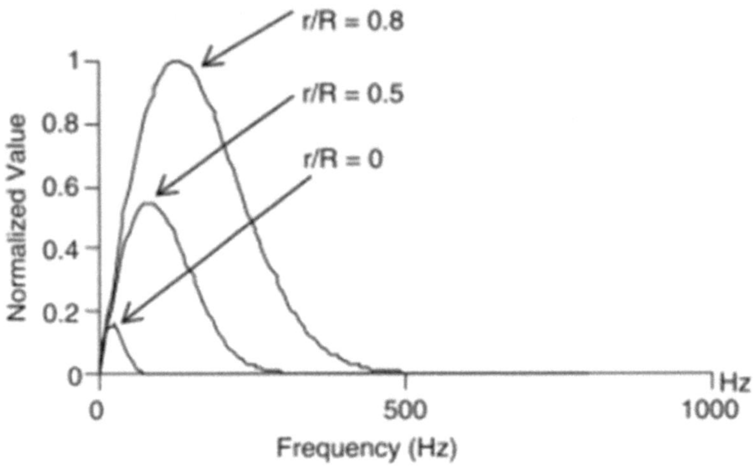

Fig. 12. Frequency Sensitivity of a circular Dynamic Electro static meter

Fig.11 provides a temporal spatial sensitivity. The corresponding frequency spatial behaviour of circular electrostatic meter is shown in Fig. 12 [24], In this figure, r is the radial coordinate from the pipe central line, R is the radius of the sensor. The vertical coordinate represents normalised output signal when a roping flow stream in parallel with pipe axial central line, passes through the sensor at different radial positions. It can be seen that the meter produces the signals with higher magnitude and wider frequency band when the stream is at r/R=0.8, compared that with the roping stream passes the central line (r/R=0). The figure does not provide the response to the roping stream passing from the location where r/R is greater than 0.8. However from Fig.11, it can be predicted, the magnitude will be lower, and the frequency band will be wider.

Theoretically, if the frequency components can be split and weighted according to where the flow stream passing through, a uniform sensitivity of meter can be achieved, which is one of possible solutions for non-uniform sensitivity compensation.

4.4 EST (Electrostatic Tomography)

Represented by Capacitance Tomography (ECT), "Process tomography" has attracted great attention since 1980s [43] [44] [45], and the research in this area has made significant progress. Besides ECT, there are many different types of tomographic techniques such as Electrical impedance tomography (EIT), optical tomography and Electrostatic Tomography [46]. "Procee tomography" uses an array of sensors mounted on the boundary of a vessel or a pipe to detect the pixel flow concentration and velocity in process. The flow profile can be reconstructed based on the information obtained from the sensor array. Theoretically, this is an ideal method to solve the problems caused by non even solids distribution in air solids two-phase flow.

As the name suggests, electrostatic tomography (EST) uses an array of electrostatic sensors to detect the distribution of charges carried by particles and particle velocities. If the amount of charge carried by particles to concentration ratio is constant, the flow rate of solids can then be derived based on the integration of the product of pixel concentration and velocity over a given cross sectional area.

For any type of process tomography, the successful realization depends on sensing system design, signal conditioning, signal to noise (S/N) ratio, proper data acquisition system and efficient algorithm.

EST is a passive sensing system, which is one of its advantages [47] over ECT. However inherently, for the same number of sensors (electrodes), the resolution of EST is lower than that of ECT. Combined systems (dual modality) [48] can offer better resolution and reliability. Fig. 13 provides the simulation results for an EST and an EST/ECT combined systems [2]. In this figure, a uniform positive charge density distribution in a stratified flow at the bottom half of a pipe is assumed, the reconstructed image using information from the EST system only in Fig.13b is vague, the boundary is not clear. Compared to the image in Fig.13b, Fig.13c offers much better result which is obtained by combining the information from the EST and ECT of dual modality system.

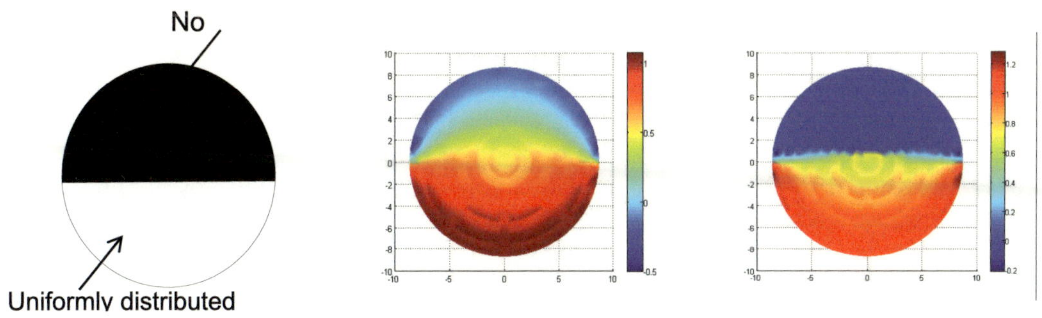

(a) (b) (c)

Fig. 13. Image Reconstruction from an EST/ECT

5. Current research in this area

At present, the modelling of charge induction with consideration of particle dielectric property is the new development in this area [2]. The research to develop an overall model to relate the signal rms, solids velocity and solids mass flow rate is under way [26]. The electrostatic method used in square pipe lines [49] has also been investigated, and the study on the effect of radial velocity on flow measurement is useful for understanding of the mechanism of electrostatic meters [31]. The technique of combing ECT and EST for gas solids flow measurement opened a new frontier.

At the time of writing this chapter, the electrostatic technique have been successful in some areas, for example in measuring flow split among pneumatic conveyors, in providing warning of blockage and for inferring primary air flow rate measurement . Some research outcomes are yet to be applied in practice. It is envisaged that the techniques will be further improved for flow measurement and flow regime diagnose not only in lean-phase conveying as in coal-fire power generation, but also for dense phase flow as in gasification and in blast furnace feeding.

6. References

[1] Dechene. R.L., Farmer A.D., ' Triboelectricity—A tool for solids flow measurement', Industrial Minerals, May 1992, pp131-135

[2] Zhou Bin, Zhang Jianyong, Wang Shimin, "Image reconstruction in electrostatic tomography using a priori knowledge from ECT", Nuclear Engineering and Design Volume 241, Issue 6, June 2011, pp 1952-1958.

[3] British Standard, BSI PD CLC/TR 50404:2003, "Electrostatics—Code of practice for the avoidance of hazards due to static electricity", Published on 9 July 2003.

[4] Batch B.A., Dalmon J. and Hignett E.T., "An Electrostatic probe for measuring the particle flux in two-phase flow", Laboratory note No. RD/L/N115/63, Central Electrcity Research Laboratories, 7 November 1963

[5] Cooper W.E., "The electrification of fluids in motion", British Journal of Physics 4 Suppl.No. 2, 1953 Couvertier. P. 1961, C.R. Akad.Sci., 252 No. 12, 1726-7

[6] Hignett, E.T. 1963, Ph.D. Thesis, University Of Liverpool.

[7] Curtis, W.D., Logan, J.D., Parker, W.A., "Dimensional analysis and the pi theorem", Linear Algebra and Its Applications, 1982, 47 (C), pp. 117-126

[8] Hignett E.T., "Particle-charge magnitudes in electrostatic precipitation", PROC. IEE, vol.114, No.9, September 1967, pp1325-1328.

[9] Soo S.L., Tung S.K., "Pipe flow of suspensions in turbulent fluid - Electrostatic and gravity effects" Applied Scientific Research, Volume 24, Issue 1, December 1971, Pages 83-97.

[10] P. W. King, 1973) "Mass flow measurement of conveyed solids by monitoring of intrinsic electrostatic noise levels", *Proc. 2nd Int. Conf. on the Pneumatic Transport of Solids in Pipes PNEUMOTRANSPORT 2*, University of Surrey, Guildford, England, pp 9-19 , 1973

[11] Bindemann K. C. G., Miller D. & Hayward P., (1999). "Kingsnorth pf flow meter demonstration trials". DTI Report Coal R167.

[12] Foster Wheeler Power Group Inc, "New ECT STAR makes On-line Coal Flow Measurement even more powerful", Design Bulletin, No 109.5 November 2002.

[13] Cole B. N., Baum M. R., Mobbs F.R., 'An investigation of electrostatic charging effects in High-speed gas-solids Pipe flows', Proc. Instn. Mech. Engres. Vol 184 , 1969-70, pp77-83.

[14] Masuda Hiroaki, Komatsu Takahiro, Mitsui Naohiro and Iinoya Koichi, "Electrification of gas-solid suspensions in Steel and insulating-coated pipes", Journal of Electrostatics, Volume 2, 1976-1977-pp341-350

[15] Yan Y. "Mass flow measurement of pneumatically conveyed solids", Ph .D thesis, University of Teesside 1992.

[16] Coulthard J., "solids flow measurement using ring-shaped electrostatic sensors",Industrial report to Davy McKee Stockton, June 1983.

[17] Gajewski J.B.and Szaynok A., "Charge measurement of dust particles in motion", Journal of Electrostatics, Volume 10, 1981, pp229-234.

[18] Gajewski J.B., Szaynok A., " Charge measurement of particles in motion, part I, J. Electrostatics, 10 (1998) pp229-234

[19] Gajewski J.B., Glòd B.J., Kala W.S., "Electrostatic method for measuring the two-phase pipe flow parameters", IEEE Transactions on Industry Applications, Volume 29, No.3, May/Jue 1993, pp 650-654.

[20] Massen R., 'Sensor with coded apertures', J. Phys. E: Sci. Instrum. Vol. 20, 1987, pp409-416

[21] Hammer E. A., Green R. G., 'The spatial filtering effect of capacitance transducer electrodes', J. Phys. E: Sci. Instrum., Vol.16, 1983, pp 438-443

[22] Cheng Ruixue, " A study of Electrostatic Pulverised Fuel Meters", Ph. D thesis, Teesside University, 1996.

[23] Zhang Jianyong, "A Study of an Electrostatic Flow Meter", Ph. D thesis, Teesside University, 2002.

[24] Zhang Jianyong, Coulthard John, "Theoretical and experimental studies of the sensitivity of an electrostatic pulverised fuel meter", Journal of Electrostatics, Vol.63, issue 12, pp 1133-1149, OCT. 2005.

[25] Zhang Jianyong, Coulthard John, "On-line Indication of Variation of Particle Size Using Electrostatic PF Meters", The proceedings of the 31st International Technical Conference on Coal Utilization & Fuel Systems, Clearwater, Florida, USA, May 21 to 25 2006.

[26] Zhang Jianyong; Cheng Ruixue; Coulthard John, "Calibration of an electrostatic flow meter for bulk solids measurement in pneumatic transportation", *Bulk Solids Handling* 28 (5), pp. 314-319, 2008

[27] Xu C. "Gas-solids Two phases Flows Parameters and Particle Electrostatic Measurement", Ph D thesis, Souheast University, Nanjing, China, 2005

[28] Xu C., Tang G., Zhou B., Yang D., Zhang J., and Wang S., "Electrostatic introduction theory based spatial filtering method for solid particle velocity measurement", AIP Conference Proceedings ,Volume 914, 2007, Pages 192-200.

[29] Zhang Jianyong; Cheng Ruixue; Al-Sulaiti Ahmed, "Velocity Measurement of Pneumatically Conveyed Solids Based on Signal Frequency Spectrum", 8th International Conference on measurement and control of Granular Materials, Shenyang, China, 27-29 Aug. 2009

[30] Laux S., Grusha J, Rosin T., Kersch J.D., "ECT: More than just coal-flow monitoring", Modern power systems, March 2002, pp 22-29.

[31] Zhang Jianyong; Zhou Bin;Wang Shimin; "Effect of axial and radial velocities on solids mass flow rate measurement", Robotics and Computer-Integrated Manufacturing, volume 26, Issue 6, December 2010, pp 576-582

[32] Gajewski J.B., "Electric charge measurement in pneumatic installations", Journal of Electrostatics, Volume 40-41, 1997, pp 231-236

[33] ECSC Project 7220-PR-050 Report: Powergen, U. Teesside, Sivus, ABBCS, IST ISTCM, ABB Kent , "Measurement & control techniques for improving combustion efficiency & reducing emissions from coal-fired plant", 2002.

[34] Beck M. S., Plaskowski A., 'How Inherent flow noise can be used to measure mass flow of granular materials', Instrument Review, November 1967, pp 458-461

[35] Coulthard J, "Ultrasonic Cross-correlation Flowmeters", Ultrasonics,Volume 11, Issue 2, 1973, March, 1973

[36] Coulthard J., Keech R.P., Asquith P., "The Electrostatic Cable meter", Inst. Of Physics International Conference on Electrostatics", York May 1995.

[37] Cheng R. , Wang S.C." Microprocessor based on-line real time cross correlator for two-phase flow measurement") Transactions of University of Science and Technology, Beijing, China.1991,13(6).-572-576, (In Chinese)

[38] Keech ray, Coulthard John, Cheng Ruixue, “Measurement using cross correlation”, Patent No WO/1998/055839, filed 06 May 1998, International Application No. PCT/GB1998/001654, publication date, 10.12.1998.

[39] Zhang Jianyong, “Project Final report (RFCS-CR-03005) On-line measurement of coal quality parameters by inference of sensor information”, for the work carried by Teesside University, Dec. 2007

[40] N. Huber and M. Sommerfield, Characterization of the cross-sectional particle concentration in pneumatic conveying system. Powder Technol., 79 (1994), pp. 91–210.

[41] Frank Th., Bernert K., Pachler K., Schneider H., “Aspects of efficient parallelization of disperse gas-particle flow prediction using Eulerian–Lagrangian approach”, Proceedings of ICMF-2001, Fourth International Conference on Multiphase Flow, May 21–June 1, 2001 New Orleans, LA, USA, paper 311.

[42] Peng B., Zhang C., Zhu J., “Numerical study of the effect of the gas and solids distributors on the uniformity of the radial solids concentration distribution in CFB risers”,Powder Technology,Volume 212, Issue 1, 15 September 2011, Pages 89-102

[43] Wang, H., Wang, X., Lu, Q., Sun, Y., Yang, W., “Imaging gas-solids distribution in cyclone inlet of circulating fluidised bed with rectangular ECT sensor”, *2011 IEEE International Conference on Imaging Systems and Techniques, IST 2011 - Proceedings*, art. no. 5962181, pp. 55-59

[44] Williams, R.A., Beck, M.S. “Process Tomography - State of the Art”, Proceedings of the first World Congress on Industrial Process Tomography, Buxton, UK , 1999, pp. 357-362.

[45] Yang W.Q. , Stott A.L , Beck M.S. , Xie C.G. , “Development of capacitance tomographic imaging systems for oil pipeline measurements”, Review of Scientific Instruments, Volume 66, Issue 8, 1995, Pages 4326-4332

[46] Green R.G., Rahmat M. F., Evans K., Goude A., Henry M., and Stone J. A. R., "Concentration profiles of dry powders in a gravity conveyor using an electrodynamic tomography system", *Meas. Sci. Technol.*, vol. 8, no. 2, pp.192 - 197 , 1997.

[47] Machida M., Kaminoyama M., “Sensor design for development of tribo-electric tomography system with increased number of sensors”, Journal of Visualization, Volume 11, Issue 4, 2008, Pages 375-385.

[48] Basarab-Horwath I., Daniels, A.T., Green, R.G. “Image analysis in dual modality tomography for material classification”, Measurement Science and Technology, Volume 12, Issue 8, August 2001, PP 1153-1156.

[49] Peng Lihui, Zhang Y. Yan Y., “Characterization of electrostatic sensors for flow measurement of particulate solids in square-shaped pneumatic conveying pipelines”, Sensors and Actuators A: Physical, Volume 141, Issue 1, 15 January 2008, Pages 59-67

Part 4

Mathematical Modelling

4

Mathematical Models for Electrostatics of Heterogeneous Media

Toshko Boev
Department of Differential Equations, University of Sofia, Sofia, Bulgaria

1. Introduction

Since most of twenty years intensive investigations, in Physical Chemistry and applied Physics, have been directed to various basic processes for complexly structured material systems, with a strong accent on surface phenomena problems. Such systems, known as heterogeneous, consist typically of different bulk phases, separated by specific interfaces, which can be also realized as containing distinct but near by sub-phases. Additionally, the common line contours, of 2D sub-phases, are treated as materially autonomous (1D) phases as well. One of the main directions in said topics consider surface nucleation phenomena in gas-liquid systems. The interest here has been provoked mainly from the open questions for the mechanism of the surface nucleation - in particular in lipid systems. Said questions essentially concern basic topics of Physical Chemistry, related also to ecological applications. As a second class, note the problems on structure building of semiconductor films of air-crystal media, via an actual technological interest: it primarily concerns the main factors governing the growth and roughness of semiconducting surface films. Another (third) class of related topics is shown by the recent studies on cell biology problems (e.g. [4]). This class includes also mathematical models for detecting of anomalies in the human organic systems - for instance, the blood circulatory system and that of the white liver oxygen transfer.

The electrostatic properties of matter have been taken as the basic framework for investigations of surface phenomena problems. Especially, adequate expressions have been sought for the electric potential - as the key quantity, integrating the basic electrostatic parameters of medium. Our main goal here consists in finding such expressions, primarily concerning the interface potential of complex (3-2-1D) heterogeneous systems. Said aim yields the key question how to construct a proper mathematical model of the matter electrostatics, which introduces a correct problem for the electric potential. Secondly, it is necessary to do the main steps in the mathematical analysis of the relevant problem. Here we propose an answer of the above question, taking into account two required basic steps. The first one consists in introducing the object called heterogeneous media, when in reality we have given a material system of different bulk (3D) phase, for instance - gas and liquid, with a relatively thin transition layer (say emulsion). In our treatment an additional stage of heterogeneity is presumed: the bulk transition layer consists also of two near-by sub-phases (of differing matter). According to the Gibbs idealizing approach ([6]), we have to consider said transition layer as a material 2D formation S - the common surface boundary of the two bulk phases. Thus the first stage of introducing heterogeneous media results in the (Gibbs idealized) heterogeneous system $\{B^- \cup S \cup B^+\}$, consisting in two bulk phases - B^+, B^-, and the 2D phase S - as an interface. The above mentioned stage (of second order heterogeneity) should be noted as a first point of new elements in our results: we assume the interface in the form $S = \{S^- \cup l \cup S^+\}$, with two 2D sub-phases - S^+, S^- (of generally differing 2D materials), and a line material component l - the common boundary of S^+, S^-. Component l is assumed homogeneous and introduced by applying again the Gibbs idealizing approach, taken now on the interface. Next, let us comment the second required basic step of modeling. Because the aim is to model electrostatics, we have to deal with the Maxwell electrostatic system, as a constituting-phenomena low, applied however for the totally heterogeneous medium $\{B^- \cup S^- \cup l \cup S^+ \cup B^+\}$. Note here the specific detail concerning the charge density ρ (which essentially enters in the Maxwell system): via an electrochemical principle, we should presume ρ depending on the electric potential u, i.e. it generally holds $\rho = \rho[u]$. Moreover, function $\rho[u]$ takes the form of the known Boltzmann distribution, for instance in case of electrolytes. Because the Gibbs idealization assumes a step transition across the interface (consequently again such transition, but of lower dimension, is assumed across the phase contour l), the next appearing problem reads: how to formalize said step transitions, in order to use effectively the Maxwell system. A useful suggestion for the first transition stage (across the interface) can be found in the monograph of D. Bedeaux and J. Vlieger ([2]). According to Bedeaux - Vlieger, we introduce the relevant material characteristics across the interface by a (first level) decomposition scheme of singularities, using Heaviside step functions $\eta^{\pm}$ regarding respectively the bulk phases $B^{\pm}$, and Dirac delta function δ_s, supported on the interface. As a second new point of our modeling, we introduce analogous decomposition scheme on the surface S, using in particular delta function supported on the contour l. After a technical procedure of solving the Maxwell electrostatic system by singular solutions, it can be established the following final form of our electrostatic model for the class of heterogeneous media $\{B^- \cup S^- \cup l \cup S^+ \cup B^+\}$:

$$\nabla^2 u = -\varepsilon_0^{-1}(\varepsilon^{\pm})^{-1}\rho^{\pm}[u] \text{ (in } B^{\pm}) \tag{1.1}$$

$$J_S[u] + \varepsilon_s^{\pm}\nabla_S^2 u = -\varepsilon_0^{-1}\rho_s^{\pm}[u] \text{ (on } S^{\pm}) \tag{1.2}$$

$$J_l[u] = -\beta_l[u] \text{ (on } l) \tag{1.3}$$

Above $u = u(x,y,z)$ is the electric potential, which is sough as bounded continuous function, regular enough in the relevant 3D and 2D phases (domains) of the material system; ∇^2 is the 3D Laplace operator and ∇_S^2 is a tangential to surface S Laplace operator; J_S is a jump type operator acting on the normal to S derivative of potential u, and, by analogy - for operator J_l, concerning contour l; $\rho^{\pm}$ and $\rho_s^{\pm}$ are the charge density terms, respectively for the bulk ($B^{\pm}$) and surface ($S^{\pm}$) phases, and β_l is an analogous quantity, for 1D phase l; $\varepsilon_0 = 8.85\, pF / m$ is the known absolute dielectric permittivity, $\varepsilon^{\pm}$ and $\varepsilon_s^{\pm}$ are the (relative) dielectric permitivities, respectively for the matter of phases $B^{\pm}$ and $S^{\pm}$, with $\varepsilon_0^{-1} = 1/\varepsilon_0$,

$(\varepsilon^{\pm})^{-1} = 1/\varepsilon^{\pm}$. Mathematically, relations (0.1) - (0.3) present a new type of transmission problem (cf. the Colton - Kress monograph, [3]). In the above generality of formulation, problem (0.1) - (0.3) remains however as open one.

In this chapter we give the main steps of deriving and solve two sub-cases of the general problem (0.1) - (0.3), related to heterogeneous systems with flat interfaces, respectively of semiconductor and organic nature. A straight line contour l enters in both the models as 1D phase of anomalies. Note that the model, with a defect line on the semiconductor interface, is closely related to real experimental data, found by scanning tunneling microscopy. On the other hand, in the case of organic interface (considered in our second model as the known *lamina basale*), the lamina folio is supposed cleft in two sub-phases by a (straight) line of functionally anomalous intercellular spaces (holes). In both the models bulk charge densities $\rho^{\pm}[u]$ are replaced with their linear approximations, however charge quantities $\rho_s^{\pm}[u]$ and $\beta_l[u]$ enter nonlinearly in the cases, respectively of semiconductor and organic interfaces. By transforming relations (0.2), (0.3), we derive and solve the relevant integral equations for the surface potential u_S. We obtain also effective formulas for certain approximations of u_S (in the case of S - a semiconductor folio), and - for the exact potential (in the case of organic S). Recall here that the contemporary problems primarily focus in detecting the surface values of the electric potential. As a consequence of finding the surface potential, we express in addition, by the classical Dirichlet problem, the bulk potentials $u^{\pm}$ as well. Thus we get expressions for potential u, valid far from the interface, which ensure in particular important diagnostic analyses (made, for instance, by parametric identification inverse problems).

In Sect. 2 we give phenomenology comments and a common derivation of the two considered models. The basic results on the surface potential are presented in Sect. 3. The question for determining of explicit approximations to $u(x,y,z)$ is discussed in Sect. 4, in case of semiconductor interface.

2. Elements of phenomenology and mathematical modeling

Let us give firstly some phenomenology comment on said two classes of heterogeneous media. Beginning with the case of organic interface, we should note the following. The interest of tools for biomedical detections of anomalies in the human circulatory system, via the walls-structure of the blood vessels, is directly motivated from the quite specific ruling function of the wall-layers. To recall and clarify the main (simplified) viewpoints here (cf. e.g. [11]), we assume a stretched location of the wall, as the flat surface ($z = 0$) on Fig. 1, below. Said construction is introduced as an admissible version of the real situation: a 3D localization is made to a capillary (practically cylindrical) vessel in the human white liver and the vessel-wall is (functionally) identified with its middle layer (called lamina basale). In reality the wall is a 3D organic threefold layer, deep not less than 120-180 nm and the midmost (just the lamina basale) is of corpulence about 40-60 nm. The upper (external) and lower (internal) layers, built - as a short description - respectively of endothelian and adventitale cells, are neglected. They are considered with a secondary role (compared to lamina basale). As known, acting as a typical bio-membrane of polysaccharide-matter, with a fine fibers structure, lamina basale is the main factor for the oxygen transfer to the blood. Via the Gibbs approach (and taking into account the ratio of the wall corpulence to the radius of the capillary vessel, which is $<<1$), we consider lamina basale layer as infinitely thin; thus we get an interface film of organic matter in an air-blood (vacuum-blood) heterogeneous 3D media. On the lamina (2D) film it is uniformly distributed a set of points – presenting the holes (tunnels, in reality radial to the vessel axis and known as intercellular spaces), which provide the oxygen contact to the blood; they are assumed however of certain functional anomaly, extremely activated on a relatively narrow (cylindrical, in reality) strip, interpreted, following Gibbs, as the middle circumference of the strip, across to the vessel-axis. Thus a specific homogeneous 1D matter phase (of the extreme anomalies) has appeared. Stretching (locally) the curved anomaly-line (and the surrounding cylindrical surface, together), we get the above mentioned (flat-interfaced) construction. Now the organic lamina-film can be presented as the plain $z=0$ (regarding a Cartesian (x,y,z)-coordinate system), where said 1D contour, of defective air permeability, has already shaped as a straight line. We shall take this line as the Oy - axis. This manner the lamina-film is cleft in two electrostatic equivalent 2D (sub-) phases by the anomaly contour and we have given a typical case of 3-2-1 D heterogeneous system, schematically shown on Fig. 1, below. The system consists in upper and lower 3D (bulk) phases, respectively of air and liquid, and a complex-structured organic interface (with a special role of a line phase). The bulk phases B^+, B^- fill the subspaces $z>0$ (B^+), $z<0$ (B^-) and their common 2D boundary – the organic interface S – is given (as already noted) by the equation $z=0$; S consists in the two neighbouring surface phases S^- ($x<0, z=0$), S^+ ($x>0, z=0$), separated by the anomaly line $l=Oy$, as an autonomous phase of 1D matter. To forecast certain influence of vessel-zones, relatively far from the phase contour l, surface phases S^-, S^+ are presumed with prescribed asymptotic values ($\varphi_\infty^-, \varphi_\infty^+$) of the electric potential.

It is possible however a sharp variant of anomalies: an air volume can leave involved between the blood and the surface, shaping an internal (lower) air bulk phase. Such presence of two-side bulk air phases is due to the anomalous air transfer: the outgoing stage gets blocked, after a previous air invasion. We would have then a heterogeneous system with upper and lower air (vacuum) bulk phases, a two-phased lamina-interface (as a 2D film) and a separating the surface phases homogeneous 1D material (straight line) phase.

Another class of heterogeneous media is that including a semiconductor interface. The model under consideration relates to electrostatics for the specific case of air-gas matter, with a semiconductor separating surface (interface), which includes moreover a defect straight line (considered below as the Oy- axis, see Fig. 2). In said case of systems the importance of the interface electrostatics is motivated, as already noted, from actual technological questions (e.g. [5], [7]).The structure of such a system can be explained as a space location of given electronic device. For a short description we will take into account the following. By a teen boundary wall of semiconductor-matter it is closed a volume of gas, and the external medium is of air. This boundary, generally curved, will be treated here as flat (observing a small part of it). Thus the system possesses two bulk phases – of internal (gas) and external (air) media, and a flat semiconducting interface, with a fine surface

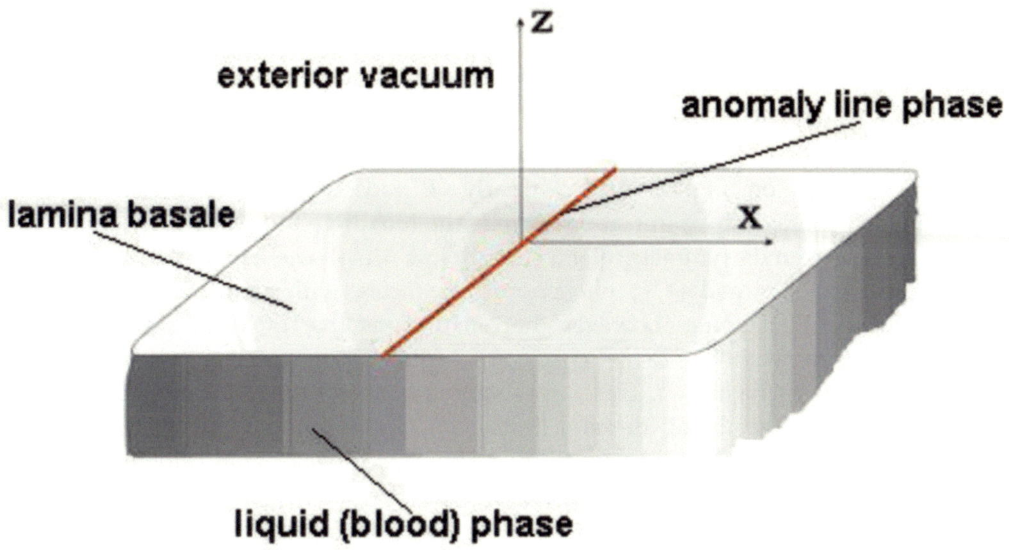
z
exterior vacuum
anomaly line phase
lamina basale
x
liquid (blood) phase

Fig. 1.

roughness, as a straight line defect. The bulk phases are considered as materially equivalent (3D) sub-domains of vacuum. The separating boundary is of indium-phosphorus, InP(110), semiconductor: the real corpulence of the InP(110)-wall is neglected and the wall is identified with its external surface film. The defect line, playing the role of a homogeneous detachment, i.e. of an electrostatic autonomous material 1D phase, separates said interface in two surface (2D) phases, denoted as before by S^{-} ($x<0, z=0$), S^{+} ($x>0, z=0$). It is posed again a typical case of 3-2-1 D heterogeneity – by a material system, interpreted as vacuum-semiconductor-vacuum (Fig. 2). The InP(110) surface film is presumed as the plain $z=0$ as well (see Fig. 2) and the Oy - axis is oriented on the defect straight line. The vacuum bulk phases fill the upper and lower semi-spaces, $z>0$ and $z<0$, respectively. Each 2D phase (on $z=0$) is characterized by an essentially dominating distribution of positively charged phosphorus vacancies, while, as a key anomaly, the line phase l ($l=Oy$) enters in the surface electrostatics symmetrically surrounded by an extremely narrow strip of width $2d$ ($d>0$). The whole this band is denuded of phosphorus vacancies ([5], [7]). The above construction is essentially supported by real experimentally found data. A credible visual result of [5] and [7] (see Fig. 2, [7]) has been found by the so-called scanning tunneling microscopy. The picture (Fig. 2, [7]) shows the surface structure, fixed after annealing of InP(110) samples at temperatures up to 480 K, followed by heat normalizing. The scanned image includes two near by surface domains (let us denote them by P^{-}, P^{+} - see the semi-planes $x>d, x<-d$ on Fig. 2, below; clearly $P^{\pm} \subset S^{\pm}$). These zones are materially equivalent and separated by a transition strip T. Its breadth ($2d$) really is less than 10 nm ([7]). The strip surrounds symmetrically a straight line l (see Oy, Fig. 2). Each of P^{-}, P^{+} is filled by positively charged (+1e) phosphorus vacancies, with about 5.5 nm ([5]) mean distance between them. Note however that strip T is free of vacancies, but remains generally charged with about +2e (per spacing of 0.6 nm, [7]) mean magnitude of charges on the axis of symmetry (the line l). This way T enters as an electrostatic autonomous surface component. On the other hand the ratio {[area](T) / [area]($P^{-} \cup T \cup P^{+}$)} is negligible to consider T as an equipollent (say to P^{-}, P^{+}) 2D surface component. Moreover, the two relations - that of T-wide (10 nm) to the above density unit (5.5 nm), and the other – of possible (averaged) electrostatic impact of T to the influence of its middle axis l, allow some identifying of the strip T with the axis l; thus l takes the role of an intrinsic 1D phase. Let us note that, via the Gibbs approach, the semi-zones ($T^{+}(x>0), T^{-}(x<0)$) of T seem to complicate additionally the surface heterogeneity, imposing – as a (Gibbs) principle – new line phases: the contours $l_d^{-}=\{x=-d; z=0\}$ and $l_d^{+}=\{x=d; z=0\}$. These (new) phases however enter also negligibly in the surface electrostatics: across l_d^{-}, l_d^{+} the surface electric field stays continuous, under equal permitivity-values ε_s. So-described picture (of a smooth, flat 2D film) represents a real surface layer with certain nanoscopic roughness, due to step defects. In reality l is actually the edge of a step, deep about 4 nm ([7], Fig. 1), P^{-} and P^{+} are the terraces (lower and upper, say) of the step, and T is a space construction, divided by l in two halves, T^{-} (lower) and T^{+} (upper), marking off the edge from the relevant terraces.

The next stage of this section is to sketch the basic step of modeling. Via the introduced framework (see Sect. 1) the key tool for description of electrostatic phenomena in complex media relates to the Maxwell system (in case of dielectrics, e.g. [10], [12]):

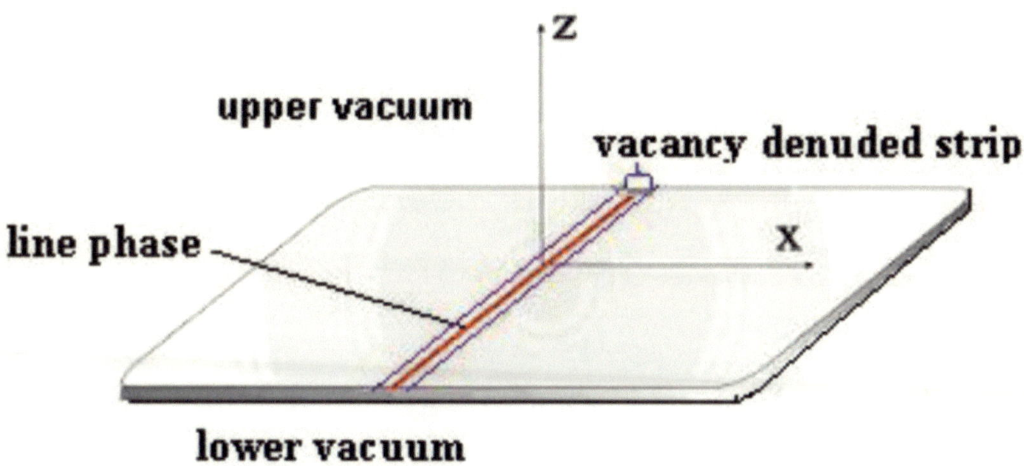

Fig. 2.

$$\text{a) } \nabla . \mathrm{D} = \rho \text{ ; b) } \mathrm{D} = -\varepsilon_0 \varepsilon \nabla u \,. \tag{2.1}$$

Here ∇ is the nabla operator, D is the vector of the electric induction ([12]), called also (in Electrochemistry, e.g. [9]) electric displacement, $\nabla . \mathrm{D}$ is the formal scalar product of the vectors nabla and D, i.e. $\nabla . \mathrm{D} = div\ \mathrm{D}$; ρ is the charge density; ε is the relative dielectric permitivity for the relevant part of the medium (in particular $\varepsilon = \varepsilon_b^-$, at $z < 0$, $\varepsilon = \varepsilon_s$, at $z = 0, x \neq 0$); u is the electric potential, $\nabla u = grad(u)$, where $(-\nabla u)$ represents the electric field, propagated in the whole 3D material system. Equations (2.1) hold for the total (3D) system and, as known, potential u is a continuous function of (x,y,z), in spite of the various material phases; the heterogeneity of the system is indicated however mainly by the quantities D and ρ (note that the permitivity ε enters in these quantities). Next, from the singular decompositions, mentioned in Sect.1, applied below for quantities D and ρ, we get the following problem. Find the (admissibly regular) solutions (D, u) to (2.1), corresponding to the said singular decompositions.

Both the considered cases of heterogeneous systems are however homogeneous on the y - direction, due to assumed homogeneity of the 1D phase l, and the electric potential $u = u(x,y,z)$ will actually depend on x,z, i.e. $u = u(x,z)$. Applying systematically a double decomposition scheme in reworking of the Maxwell system (see below), we shall establish the following final formulation to the sought mathematical models:

$$\nabla^2 u = \kappa_b^2 u \;\; (z \neq 0),\, x \in R^1 \,; \tag{2.2}$$

$$|\, u \,| \leq const.,\, (x,z) \in R^2 \,; \tag{2.3}$$

$$u(x,+0) = u(x,-0),\;\; x \in R^1 \,; \tag{2.4a}$$

$$\Delta[\varepsilon_b u_z] + \varepsilon_s u_{xx} = \varepsilon_s k_s^2 (u - \varphi_\infty) + \rho_s^*[u]\,,\;\; x \neq 0\,; \tag{2.4b}$$

$$u(-\infty,0) = \varphi_\infty^-,\;\; u(+\infty,0) = \varphi_\infty^+ \,; \tag{2.5}$$

$$u(-0,0) = u(+0,0)\,; \tag{2.6a}$$

$$\varepsilon_s^+ u_x(+0,0) - \varepsilon_s^- u_x(-0,0) = -\beta_l[u]\,. \tag{2.6b}$$

In (2.2) $\nabla^2 \equiv \partial_x^2 + \partial_z^2$ is the Laplace operator; u_x, u_z, u_{xx} are first or second order derivatives regarding the relevant variable; $u(x,+0)$, $u(x,-0)$ are respectively the limits (supposed finite) $\lim\limits_{z\to 0} u(x,z)$ (at $z > 0$ or $z < 0$), and, by analogy - for $u_z(x,+0)$, $u_z(x,-0)$; $u(+0,0) = \lim\limits_{x\to 0} u(x,0)$ and $u_x(\pm 0,0) = \lim\limits_{x\to 0} u_x(x,0)$, respectively at $x > 0, x < 0$, both - for $u(\pm 0,0)$ and $u_x(\pm 0,0)$; $u(\pm\infty,0) = \lim\limits_{x\to\pm\infty} u(x,0)$. (Above R^m is the real m - dimensional Euclidean space, $m = 1,2,\ldots$.) As known, parameters k_b, ε_b and k_s, ε_s are the main factors of the system-electrostatic nature; they are given step constants: $\varepsilon_b = \varepsilon_b^+(z > 0)$, $\varepsilon_b = \varepsilon_b^-(z < 0)$, $\varepsilon_b^+, \varepsilon_b^-$ - positive (and generally different); $k_b = k_b^+(z > 0)$, $k_b = k_b^-(z < 0)$ are nonnegative constants in (2.2); $k_s = k_s^+(x > 0)$, $k_s = k_s^-(x < 0)$ - in (2.4), with positive k_s^+, k_s^-; by analogy, $\varepsilon_s = \varepsilon_s^+(x > 0)$, $\varepsilon_s = \varepsilon_s^-(x < 0)$ - in (2.4), (2.6), with $\varepsilon_s^+ > 0$, $\varepsilon_s^- > 0$ - constants. The material meaning of parameter k_s (by analogy from that of k_b) is expressed by the quantity $k_s^{-1} = \dfrac{1}{k_s}$, known as the surface Debye length (e.g. [13], or the surface screening length (e.g. [7]). Parameters ε_b, ε_s are respectively the bulk and surface dielectric permitivities, with $\varepsilon_b^+(\varepsilon_b^-)$, $\varepsilon_s^+(\varepsilon_s^-)$ - for the relevant bulk and surface phases. The asymptotic values of the potential are prescribed by the parameter φ_∞ (a given quantity): $\varphi_\infty = \varphi_\infty^\pm$ (at $x>0, x<0$), where $\varphi_\infty^+, \varphi_\infty^-$ are real, generally different constants; the parameter $\beta_l = \beta_l[u]$ enters as $\beta_l = \frac{\rho_l}{\varepsilon_0}$ by $\rho_l = \rho_l[u]$ – the electric charge density (supposed depending on potential u) upon the line phase; $\varepsilon_0 = 8.85\, pF/m$ is the mentioned absolute dielectric permitivity. In equation (2.4.b) $\Delta[\varepsilon_b u_z] = \Delta[\varepsilon_b u_z](x,0)$ is the space-jump operator, $\Delta[\varepsilon_b u_z](x,0) \equiv \varepsilon_b^+ u_z(x,+0) - \varepsilon_b^- u_z(x,-0)$ and in the right hand side of (2.4) we have $\rho_s[u] = \varepsilon_0(\varepsilon_s k_s^2(u-\varphi_\infty) + \rho_s^*[u])$. In the case of organic interface we shall suppose $\rho_s^*[u]=0$, while for semiconductors we shall use $\rho_s^*[u] \equiv \omega_d^0(\varepsilon_s k_s^2 \frac{\Delta\varphi_\infty}{2} sg - q_s[u])$. Here $\omega_d^0 = \omega_d^0(x) \equiv \frac{1}{2d}\omega^0\left(\frac{x}{d}\right)$ is the rescaled characteristic function of the unit interval, i.e. $\omega^0(x) \equiv 1, |x|<1; \omega^0(x) \equiv 0, |x|>1$; $\Delta\varphi_\infty \equiv \varphi_\infty^+ - \varphi_\infty^-$ is the asymptotic surface power; $sg = sg(x) = \frac{x}{|x|}$ is the sign function; $q_s[u] \equiv \frac{1}{2!}[\varepsilon_s k_s^2(u-\varphi_\infty^*)]^2 - \frac{1}{3!}[\varepsilon_s k_s^2(u-\varphi_\infty^*)]^3$, where $\varphi_\infty^* = \frac{\varphi_\infty^+ + \varphi_\infty^-}{2}$. The cubic nonlinear charge density (of the surface $z=0$) is preferred just on the strip $|x|<d$ – by the rest term $\omega_d^0 q_s[u]$, while a linear approximation (regarding potential u) is assumed adequate to reality out of the strip.

Via the phenomenology-essence potential u will be searched for a bounded function (condition (2.3)), continuous in R^3, classically regular in the sets $z \neq 0$ and $x \neq 0\ (z=0)$, with continuous gradients u_z, u_x, respectively at $z \geq 0$, $z \leq 0$ (for u_z), and $x \leq 0, x \geq 0$ $(z=0)$ – for u_x. Now we can define the needed space of regular solutions to (2.2) – (2.6).

Definition. A function $u(x,z)$, with the above noted regularity, shall be called *classical solution* to problem (2.2) – (2.6) if satisfies the additional property $u(x,0) - \varphi_\infty \in L_2$ and relations (2.2) – (2.6). (L_2 is the well known space of the squared-integrable functions.)

From the vacuum assumption for the upper (external) air phase ($z>0$) we suppose from now on:

$$\varepsilon_b^+ = 1,\ k_b^+ = 0. \tag{2.7}$$

The central results below relate to determination of the surface potential (possibly by an explicit approximation) – as the key first step in solving the full (2.2) – (2.6) – problem.

Let us now sketch the main steps for derivation of the final mathematical models, starting from the Maxwell system (2.1). Via the presumed complex heterogeneity, we shall seek solutions (D, u) of system (2.1) by decompositions in two levels (bulk and surface), of the following type:

$$\text{a) } \rho = \rho_b^- \eta^- + \rho_b^+ \eta^+ + \rho_s \delta(z)\,;\ \text{b) } \rho_s = \rho_s^- \eta_s^- + \rho_s^+ \eta_s^+ + \rho_l \delta_l\,; \tag{2.8}$$

$$\mathrm{D} = (\mathbf{D}_b)^- \eta^-(z) + (\mathbf{D}_b)^+ \eta^+(z) + \mathrm{D}_s\,\delta(z)\,, \tag{2.9}$$

$$\mathrm{D}_s = (\mathbf{D}_s)^- \eta_s^- + (\mathbf{D}_s)^+ \eta_s^+ + \mathrm{D}_l\,\delta_l\,. \tag{2.10}$$

In the above relations $\eta^+(z)\,/\,\eta^-(z)$ are respectively the Heaviside forward/backward functions (i.e. $\eta^+(z)=1$, at $z>0$, $\eta^+(z)=0$, at $z<0$, $\eta^-(z)\equiv\eta^+(-z)$) and $\delta(z)$ is the Dirac delta-function, supported at $z=0$; $\eta_s^-=1$, at $z=0, x<0$ and $\eta_s^-=0$, at $z=0, x>0$, by analogy: $\eta_s^+=1$, at $z=0, x>0$ and $\eta_s^+=0$, at $z=0, x<0$; next, δ_l is delta-function, supported on the line $l: x=0\,(z=0)$, and we shall also use the notation $\delta(x)$, for δ_l. Relations (2.8.a), (2.8.b) and (2.9), (2.10) just illustrate, respectively for the charge density and the electric induction, the essential generalization, in two levels (see [13]), of the Bedeoux-Vlieger ([2]) step formalism to the bulk-surface-bulk transitions. *Remark:* terms like $\rho_b^{\pm}\eta^{\pm}(z)$ do not enter in the right hand side of (2.8.a) in the case of semiconductor interface because of the vacuum hypothesis ($\rho_b^-=\rho_b^+=0$). In (2.9) $(\mathbf{D}_b)^-$ and $(\mathbf{D}_b)^+$ are at least smooth (vector) functions of (x,z), respectively at $z<0$ and $z>0$, with finite but generally different limit values at $z\to 0$, $\forall$ fixed x and D_s is a vector function of x, assumed in the form of (2.10). Analogous presumptions hold to $(\mathbf{D}_s)^-$, $(\mathbf{D}_s)^+$ – in (2.10), as functions actually of x (with finite and different limit values at $x\to 0$), and to (scalar) functions $\rho_s^{\pm}=\rho_s^{\pm}(x)$ (considered as at least continuous respectively at $x\le 0$ and $x\ge 0$); ρ_l and D_l enter in (2.8.b) and (2.10) respectively as constant scalar and vector. Substituting from (2.8) – (2.10) into electrostatic equations (2.1), we get (with $\varepsilon_b^+=1$):

$$\nabla.\,(\mathbf{D}_b)^+ = 0\ (z>0),\ \nabla.\,(\mathbf{D}_b)^- = \rho_b^-[u]\ (z<0); \tag{2.11a}$$

$$(\mathbf{D}_b)^+ = -\varepsilon_0\varepsilon_b^+.\nabla u\,(z>0),\ (\mathbf{D}_b)^- = -\varepsilon_0\varepsilon_b^-.\nabla u\,(z<0); \tag{2.11b}$$

$$D_+^z(x,0) - D_-^z(x,0) + \nabla_s.\,\mathrm{D}_s = \rho_s[u]\ (z=0, x\ne 0); \tag{2.12.a}$$

$$\mathrm{D}_s = -\varepsilon_0\varepsilon_s.\nabla_s u\,(z=0, x\ne 0); \tag{2.12.b}$$

$$(\mathbf{D}_s)^{x,+} - (\mathbf{D}_s)^{x,-} = \rho_l\,. \tag{2.13}$$

Here we have denoted by ∇_s the tangential (to $z=0$) component of the nabla operator ∇; $D_+^z(x,0)=\lim_{z\to+0} D^z(x,z)$, and, by analogy – for $D_-^z(x,0)$, where D^z is the normal to $z=0$ component of vector D, and the limits are supposed finite, $\forall x$; $(\mathbf{D}_s)^{x,-}, (\mathbf{D}_s)^{x,+}$ are the relevant limits (also assumed finite), at $x\ne 0$, for the normal to l component $(\mathbf{D}_s)^l$ of D_s. Let us note (calculating the results of said substituting) that the normal to $z=0$ component D_s^z of vector D_s is found to vanish ($D_s^z=0$), i.e. D_s presents a flat (planar) vector field (see e.g. [13] for details). *Remark:* the used derivations of the Heaviside and Dirac delta-functions, necessary to get system (2.11) – (2.13), are taken in the Schwartz distributions meaning (e.g. [8]).

Now we will discuss the charge density terms $\rho_b^-[u]$, $\rho_s[u]$, respectively in (2.11.a), (2.12.a), especially that of ρ_s. For the vacuum-semiconductor-vacuum systems we have to take $\rho_b^-[u]=0$ (from the vacuum hypothesis). For vacuum-lamina-liquid systems it is established certain relation $\rho = \rho[u]$. Such type of dependence is well known for electrolytes by the Gouy-Chapmann theory, where $\rho[u]$ is expressed by the so-called Boltzmann distribution (see e.g. [9]) for the bulk phases, and in the case of organic interface we can replace in (2.11.a) $\rho[u]$ with its linear approximation $-\varepsilon_0\varepsilon_b k_b^2 u$. Via the real phenomena, the surface charges should also depend on the space variables by potential u, i.e. $\rho_s = \rho_s[u]$. A preliminary motivation to do that follows from the argument that the polysaharide matter of the interface admits to consider it as a lipid medium, where the potential-magnitude can be assumed relatively smaller than the basic ratio $(RT_0)/F$, which yields that linear approximations become acceptable (F, R, T_0 are – as follows – the so-called Faraday and gas constants, and the absolute temperature). The Boltzmann principle, applied for surfaces, suggests dependence $\rho_s = \rho_s[u - \varphi_\infty]$ to the surface phases; i.e. we can take again the relevant linear approximation $-\varepsilon_0\varepsilon_s k_s^2(u - \varphi_\infty)$ instead of $\rho_s[u - \varphi_\infty]$. In the case of semiconductors however a nonlinear dependence $u \to \rho_s[u]$ could be derived from a parametric expression, known as Fermi-Dirac integral (e.g. [1]). Said dependence is of exponential type regarding the potential and relates well enough to the simpler one, $\rho = \varepsilon_0 \exp(-\varepsilon k^2 u)$, used for the so-called screened Coulomb potential in the bulk phases (see [5], [7] and the literature therein). It is important that the same expression has been experimentally examined in [5], [7] (with ε_s, k_s instead respectively of ε, k), to the analysis for the surface density of phosphorus vacancies. From the above-noted viewpoint we shall chose a truncation of exponential dependence for the surface phases in the following form (taking into account the total electro-neutrality of the considered material system and the specific inclusion of component T):

$$\rho_s = \varepsilon_0\left(-\varepsilon_s k_s^2(u - \varphi_\infty) - \omega_d^0 \varepsilon_s k_s^2 \frac{\Delta\varphi_\infty}{2} sg + \omega_d^0 q_s[u]\right). \tag{2.14}$$

To get the above expression for ρ_s (in said framework) we start from the following type of exponential dependence for the surface phases: $\rho_s = \varepsilon_0\left(\exp[-\varepsilon_s k_s^2(u - \varphi_\infty)] - 1\right)$. Under the behaviour $\rho_s[t] \sim t$, at $t \approx 0$, such an expression takes into account the total electro-neutrality of the considered material system, via the upper and lower vacuum phases (by analogy to the case of gas-lamina-liquid media). The difference $u - \varphi_\infty$ is present in the exponential term from the assumption to have given asymptotic values φ_∞ of the surface potential, different from zero (far from the specific edge l). To forecast the more complicated impact of the vacancy-denuded zone $T^- \cup T^+$, with $T^- = \{-d < x < 0; z = 0\}$ and $T^+ = \{0 < x < d; z = 0\}$, we introduce the modified dependence:

$$\rho_s = \varepsilon_0\{(1 - \omega_d^0)\left(\exp[-\varepsilon_s k_s^2(u - \varphi_\infty)] - 1\right) + \omega_d^0\left(\exp[-\varepsilon_s k_s^2(u - \varphi_\infty^*)] - 1\right)\}. \tag{2.15}$$

When rewrite the above as $\rho_s = (1 - \omega_d^0)\rho_s^0 + \omega_d^0 \rho_s^1$ (for the sake of shortness), we shall deal with the linear approximation of density ρ_s^0, instead. For undertaking that, the basic motivation issues from the observation on the interaction energy between phosphorus vacancies on the surface (see [7]) – this energy seems to be relatively small. (It has been estimated in [7] with a maximal value of 65 ± 15 meV at a vacancy separation of 1.2 nm.) On the other hand we cut off the infinite exponential sum for ρ_s^1 up to the cubic term, assuming secondary the impact of the higher powers. Thus we shall presume in relation (2.12.a) the given one in (2.14), for the surface charge density of semiconductors. On the linear (1D) phase, the contour $l = Oy$, we assume $\rho_l = \varepsilon_0 \beta_l$, with β_l - given constant, for the semiconductor case, while in the organic case we prefer a nonlinear Boltzmann type model $\rho_l = \rho_l[u]$, forecasting possible unknown complications, close to the line contour.

The next main step of modeling consists in some reworking to system (2.11) – (2.13). By the right hand sides from (2.11.b) we firstly express $(\mathbf{D}_b)^+$, $(\mathbf{D}_b)^-$ in (2.11.a) and come to the Helmholtz–Laplace equations from (2.2). As noted, condition (2.3) corresponds to the physical nature of the potential (to be a space-bounded and continuous quantity). Going to the next relations, (2.4.a), (2.6.a), they show that potential stays continuous across the transition surfaces and lines. On the other hand condition (2.5) introduces the asymptotic value of the surface potential $u(x,0)$ - they are considered as experimentally known (gauged) data. Afterwards we replace D_s in (2.12.a) by the right hand side of (2.12.b) and use that $D_+^z(x,0) = -\varepsilon_0 \varepsilon_b^+ u_z(x,+0)$, $D_-^z(x,0) = -\varepsilon_0 \varepsilon_b^- u_z(x,-0)$ (with $\varepsilon_b^+ = 1$, see (2.7)). In addition we rearrange the right hand side of (2.12.a) respectively by the nonlinear density (2.8) or the linear expression $-\varepsilon_0 \varepsilon_s k_s^2 (u - \varphi_\infty)$. This way we get, from (2.12), the complicated jump condition (2.4.b). For the second jump-condition on the electric field (see (2.6.b)), it is enough to recall that $\rho_l = \varepsilon_0 \beta_l$ and $(\mathbf{D}_s)^{x,\pm} = -\varepsilon_0 \varepsilon_s u_x(\pm 0, 0)$.

This completes the sketch of derivation to final form (2.2) - (2.6) of our mathematical models.

3. The basic integral equation and finding of surface potentials

We shall reduce in this section problem (2.2) – (2.6) to a corresponding (nonlinear) integral equation – as a background for finding of explicit type presentations to the surface electric potential. Recall firstly the supposed electrostatic equivalence of the surface phases, which yields that $\varepsilon_s^- = \varepsilon_s^+ = \varepsilon_s$, $k_s^- = k_s^+ = k_s$.

For the needed technical reworks the x - Fourier transformation is systematically taken into account below – by well known conventional expressions (e.g. as in [8]). By the x -Fourier transformation to the relations in (2.2) we find ordinary differential equations (regarding z), which yield the following presentation (for a classical solution $u = u(x,z)$ to problem (2.2) – (2.6)):

$$\hat{u}(\xi,z) = \hat{\varphi}(\xi)\exp(-z|\xi|),\ z > 0\ ;\ \hat{u}(\xi,z) = \hat{\varphi}(\xi)\exp(z\sqrt{\xi^2 + \kappa_b^2}),\ z < 0\ . \qquad (3.1)$$

It is denoted here by $\hat{u}(\xi,z)$ the (partial) Fourier transformation of $u(x,z)$ - with respect to x.

In (3.1) $\varphi(x) = u(x,0)$ and $\hat{\varphi}$ is the Fourier image of φ. The jump term in (2.4.b) can be then expressed in the next form:

$$u_z(x,+0) - u_z(x,-0) = L[\varphi]\ ;\ \hat{L}[\varphi](\xi) = -\lambda(\xi)\hat{\varphi}(\xi)\ ,\ \lambda(\xi) = |\xi| + \varepsilon_b\sqrt{\xi^2 + k_b^2}\ . \qquad (3.2)$$

(Above $\hat{L}[\varphi]$ is the Fourier image of $L[\varphi]$ and $\varepsilon_b = \varepsilon_b^-$, $k_b = k_b^-$.) Thus said jump term is presented as a linear operator $L: \varphi \to L[\varphi]$, acting from $L_2(R^1)$ into the Sobolev space $H^{-1}(R^1)$ (we refer e.g. to [8], for the H^k-spaces of Sobolev). It admits to separate the key part (2.4) – (2.6) from the full system (2.2) – (2.6) in an autonomous *boundary transmission problem*:

$$L[\varphi] + \varepsilon_s \varphi'' = \varepsilon_s k_s^2 (\varphi - \varphi_\infty) + \rho_s^*[\varphi],\ x \neq 0\ ; \tag{3.3a}$$

$$\varphi(\pm\infty) = \varphi_\infty^\pm,\ \varphi(+0) = \varphi(-0)\ ; \tag{3.3b}$$

$$\varepsilon_s[\varphi'(+0) - \varphi'(-0)] = -\beta_l\,. \tag{3.3c}$$

We have denoted by φ' and φ'' respectively the first and second derivative of $\varphi(x)$, and by $\varphi(\pm\infty)$, $\varphi(+0)$, $\varphi(-0)$, $\varphi'(+0)$, $\varphi'(-0)$ – the relevant limits. Taking the substitution $\psi(x) \equiv \varphi(x) - \varphi_\infty$, the problem (3.3) reduces into a simpler one for ψ. Let us express firstly the quantity $L[\varphi]$ by $L[\psi]$; using notations $\overset{*}{\varphi}_\infty = \dfrac{\varphi_\infty^+ + \varphi_\infty^-}{2}$ and $\Delta\varphi_\infty = \varphi_\infty^+ - \varphi_\infty^-$, we get $L[\varphi] = L[\psi] + \overset{*}{\varphi}_\infty . L[1] + \dfrac{\Delta\varphi_\infty}{2} . L[sg]$. It is directly seen that $L[sg](x) = 2\sigma_0(x)$, where $\sigma_0(x)$: $\hat{\sigma}_0(\xi) = i.sg(\xi)$ $\left(i = \sqrt{-1}\right)$, and $L[1] = 0$. Now, from (3.3), we get the next reduced problem for auxiliary function ψ:

$$\psi'' - k_s^2 \psi = -\frac{1}{\varepsilon_s}(L[\psi] + \Delta\varphi_\infty . \sigma_0 - \rho_s^*),\ x \neq 0\ \ (\rho_s^* = \rho_s^*[\psi + \varphi_\infty]); \tag{3.4}$$

$$\varepsilon_s[\psi'(+0) - \psi'(-0)] = -\beta_l\,. \tag{3.5}$$

The posed ψ-problem ((3.4)-(3.5)) is considered on the space of the real functions ψ, which are continuous in $(-\infty,-0]$, $[+0,+\infty)$, tend to zero, at $|x| \to \infty$ and belong to $L_2(R^1)$; in addition they are assumed to have the classical regularity at $x \neq 0$, with finite values of the limits $\psi'(\pm 0)$. To our next step, observe before that, given a solution ψ of (3.4) – from the said class, we actually have a suitably regular and bounded solution to the equation:

$$w'' - k_s^2 w = -\frac{1}{\varepsilon_s} F_s[\psi],\ x \neq 0\ \ (F_s[\psi] \equiv L[\psi] + \Delta\varphi_\infty . \sigma_0 - \rho_s^*). \tag{3.6}$$

Multiplying the Fourier image of $F_s[\psi]$ by factor $(\xi^2 + k_s^2)^{-1}$ we find a single bounded solution of (3.6) (see below for some details). Denote this solution by $\dfrac{1}{\varepsilon_s} U_s[\psi]$. Then from the general formula for the (bounded) solutions of (3.6), we can directly get a presentation in the form: $\psi(x) = c.\exp(-k_s |x|) + \dfrac{1}{\varepsilon_s} U_s[\psi]$ (at $x \neq 0$), with a constant c. To clarify the structure of $U_s[\psi]$, let us introduce the following auxiliary functions, related to the relevant components of operator $F_s[\psi]$:

$$g(x) = -\frac{1}{\pi}\int_0^\infty \frac{\sin(x\xi)}{k_s^2+\xi^2}d\xi \quad ; \tag{3.7}$$

$$W_L[\psi](x) = \frac{1}{2k_s}L[\psi]*\exp(-k_s\,|\cdot|)(x)\,; \tag{3.8}$$

$$R_s^\infty[\psi](x) = \frac{1}{2k_s}\rho_s^*[\psi+\varphi_\infty]*\exp(-k_s\,|\cdot|)(x)\,. \tag{3.9}$$

Above $F*\Phi$ is the convolution of two (Schwartz) distributions, F and Φ (see e.g. [8]), and $W_L[\psi](x)$ is a bounded function. In addition $R_s^\infty[\psi](x)$ is also a bounded function (because $(\omega_d^0 sg)(x)$ and $(\omega_d^0 q_s[\psi+\varphi_\infty])(x)$ are compactly supported, while $g(x)$ is evidently bounded. Now, for $U_s[\psi]$, it can be found: $U_s = W_L[\psi]+\Delta\varphi_\infty.g - R_s^\infty[\psi]$ and therefore function ψ (the solution of (3.4)) satisfies the equation:

$$\psi = c.\exp(-k_s\,|.|) + \frac{1}{\varepsilon_s}\left(W_L[\psi]+\Delta\varphi_\infty.g - R_s^\infty[\psi]\right). \tag{3.10}$$

It can be easily seen that the Schwartz derivative of $U_s[\psi]$ is in $L_2(R^1)$ and that of $c\exp(-k_s\,|x|)$ is in $H^{-1}(R^1)$, belonging in addition to $L_2(x<0)$, $L_2(x>0)$. Consequently, differentiating in (3.10), we conclude that $\psi' \in H^{-1}(R^1)$ and $\psi' \in L_2(x \neq 0)$. This yields the next distribution-relation: $\psi'(x) = \psi_1(x) - \Delta\varphi_\infty\delta(x)$ (where $\psi_1 \in L_2(R^1)$ and $\delta(x)$ is the Dirac-function). Then $(W_L[\psi])'(x) = W_L[\psi'](x) = W_L[\psi_1](x) - \Delta\varphi_\infty W_L[\delta](x)$. On the other hand - for the rest components of $U_s[\psi]$ - it is not difficult to get as follows: $g'(x) = W_L[\delta](x)$; $(R_s^\infty[\psi])'(+0) = (R_s^\infty[\psi])'(-0)$. Thus we find, for derivative ψ' (at $x \neq 0$):

$$\psi'(x) = -ck_s sg(x)\exp(-k_s\,|x|) + \frac{1}{\varepsilon_s}\left(W_L[\psi_1](x) + (R_s^\infty[\psi])'(x)\right). \tag{3.11}$$

However it also holds $W_L[\psi_1](+0) = W_L[\psi_1](-0)$ and, substituting from (3.11) in the jump relation (3.5), we determine the free constant c, as $c = c_s$, with

$$c_s = \frac{1}{2}\left(\frac{\beta_l}{\varepsilon_s k_s} - \Delta\varphi_\infty.sg(x)\right). \tag{3.12}$$

In order to modify (3.10) into an integral equation regarding the surface potential φ, we introduce also the next two functions:

$$\psi_s^0(x) = \frac{1}{\pi}\int_0^\infty \frac{\cos(x\xi)d\xi}{\lambda(\xi)+\varepsilon_s(k_s^2+\xi^2)} \;;\; \psi_s^*(x) = -\frac{1}{\pi}\int_0^\infty \frac{\sin(x\xi)d\xi}{\lambda(\xi)+\varepsilon_s(k_s^2+\xi^2)}\;. \tag{3.13}$$

Now, going back to (3.10), with $c = c_s$ (see (3.12)), via formulas (3.7) - (3.9) and relation $\varphi = \psi + \varphi_\infty$, we shall obtain the basic integral equation for surface potential φ. As a preliminary step we apply the inverse operator of $I - \frac{1}{\varepsilon_s}W_L$ (I - the identity) to equation

(3.10), using that $\varepsilon_s(k_s^2+\xi^2)[\lambda(\xi)+\varepsilon_s(k_s^2+\xi^2)]^{-1}$ is the Fourier transform of the inverse operator. Reworking this way (3.10) we get the following expression, via the functions from (3.13):

$$\varphi-\varphi_\infty+\rho_s^*[\varphi]*\psi_s^0=\beta_l[\varphi]\psi_s^0+\varepsilon_s\Delta\varphi_\infty.\psi_s^{0,1}+\Delta\varphi_\infty.\psi_s^* . \quad (3.14)$$

Above $\psi_s^{0,1}=\psi_s^{0,1}(x)$ is the (Schwartz) first order derivative of function ψ_s^0 (playing a key role, together with ψ_s^*, for the behaviour of potential φ). Note that ψ_s^0 is the (unique) solution of the linear canonical version of problem (3.4), (3.5) (with $\Delta\varphi_\infty=0, \beta_l=1$); this admits, differentiating as before (3.10) (at $\Delta\varphi_\infty=0$, with ψ_s^0 instead of ψ) to use - for the analysis of $\psi_s^{0,1}$ - the integral relation:

$$\psi_s^{0,1}(x)=\frac{1}{\varepsilon_s}\left(W_L[\psi_s^{0,1}](x)-\frac{sg(x)}{2}\exp(-k_s|x|)\right). \quad (3.15)$$

(In particular (3.15) yields the finite limits $\varepsilon_s\psi_s^{0,1}(+0)=-\frac{1}{2}$ and $\varepsilon_s\psi_s^{0,1}(-0)=\frac{1}{2}$.)

Equation (3.14) is the sought basic integral equation related to problem (2.2) - (2.6). In the case of vacuum - liquid heterogeneous system, with organic interface, (3.14) takes the specialized form (with $\rho_s^*[\varphi]=0$):

$$\varphi-\varphi_\infty=\beta_l[\varphi]\psi_s^0+\varepsilon_s\Delta\varphi_\infty.\psi_s^{0,1}+\Delta\varphi_\infty.\psi_s^* . \quad (3.16)$$

We shall study the above equation for nonlinear functions $\beta_l[t]=\alpha_l t^3$, with coefficient $\alpha_l>0$. The main results for said class of charge densities are summarized in the following assertion. Below we shall use the quantity $p_{s,0}=\psi_s^0(0)$, i.e. $p_{s,0}=\frac{1}{\pi}\int_0^\infty\frac{d\xi}{\lambda(\xi)+\varepsilon_s(k_s^2+\xi^2)}$.

3.1 Proposition

For arbitrary non zero asymptotic mean value φ_∞^* of the surface potential, arbitrary parameters $\varepsilon_b^->0, k_b^-\geq 0, \varepsilon_s>0, k_s>0$, and coefficient α_l: $\alpha_l>\frac{4}{27}p_{s,0}^{-1}|\varphi_\infty^*|^{-2}$, there exists a unique continuous bounded potential $\varphi(x)$, satisfying (3.16), such that $\varphi_\infty^*\varphi(0)<0$, determined by the formula

$$\varphi(x)=\varphi_\infty+\beta_l[t_0]\psi_s^0(x)+\varepsilon_s\Delta\varphi_\infty.\psi_s^{0,1}(x)+\Delta\varphi_\infty.\psi_s^*(x)\,,\ x\neq 0\,; \quad (3.17)$$

t_0 is either the positive (at $\varphi_\infty^*<0$) or negative (at $\varphi_\infty^*>0$) root of the equation $p_{s,0}\alpha_l t^3-t+\varphi_\infty^*=0$. In addition, the relevant space potential $u(x,z)$ (with $u(x,0)=\varphi(x)$)is determined as the unique (classical) solution of problem (2.2) - (2.6), by the next formulas:

$$u(x,z)=\frac{1}{\pi}\int_{-\infty}^{+\infty}\varphi(x-t)\frac{z}{z^2+t^2}dt\,,\ z>0\,,\ x\in R^1\,; \quad (3.18)$$

$$u(x,z)=\frac{1}{\pi}\int_{-\infty}^{+\infty}\varphi(x-t)\frac{(-1)z}{z^2+t^2}K_1^0(\kappa_b\sqrt{z^2+t^2})dt\,,\; z<0\,,\; x\in R^1\,. \tag{3.19}$$

(Above $K_1^0(x)\equiv\pi xK_1(x)$, where $K_1(x)$ is the McDonald function.)

Proof. Suppose $\varphi(x)$ is a real, continuous solution of (3.16) and let for instance $x\to+0$ (in (3.16)), using that $\varepsilon_s\psi_s^{0,1}(+0)=-\frac{1}{2}$ (see (3.15)). Because $\beta_l[\varphi]=\beta_l[\varphi(0)]$, we get then relation $\psi_s^0(0)\alpha_l\varphi^3(0)-\varphi(0)+\varphi_\infty^*=0$, i.e. $t=\varphi(0)$ is (by necessity) a real solution of the algebraic equation $p_{s,0}\alpha_l t^3-t+\varphi_\infty^*=0$. Assumptions $\varphi_\infty^*\neq 0$, $\alpha_l>\frac{4}{27}p_{s,0}^{-1}|\varphi_\infty^*|^{-2}$ easily yields existence of a unique positive root $t_0=t_0^+$ of said equation (when $\varphi_\infty^*<0$), and the same for the negative one, $t_0=t_0^-$ (when $\varphi_\infty^*>0$). Conversely, let for instance $\varphi_\infty^*<0$ and take in (3.17) $t_0=t_0^+$. Function $\varphi(x)$ given now by formula (3.17) is bounded and continuous on R^1 (which is not difficult to be verified) and, letting $x\to+0$ (in (3.17)), we find $\varphi(0)=\varphi_\infty^*+p_{s,0}\alpha_l t_0^3$; i.e. $\varphi(0)=t_0$, $\beta_l[t_0]=\beta_l[\varphi(0)]=\beta_l[\varphi]$, and (3.17) shows that $\varphi(x)$ satisfies integral equation (3.16). Having the surface values $u(x,0)=\varphi(x)$, it remains to solve the following two Dirichlet problems (as already noted in Sect.1, above): $\{\nabla^2u=0, z>0; u(x,0)=\varphi(x), x\in R^1\}$ and $\{\nabla^2u=k_b^-u, z<0; u(x,0)=\varphi(x), x\in R^1\}$. As it generally known, the relevant solutions are determined respectively by (3.18), (3.19).

Consider now the case of vacuum – vacuum heterogeneous system, with a semiconductor interface; then (3.14) is written as:

$$\varphi-\varphi_\infty+\rho_s^*[\varphi]*\psi_s^0=\beta_l\psi_s^0+\varepsilon_s\Delta\varphi_\infty.\psi_s^{0,1}+\Delta\varphi_\infty.\psi_s^*\,; \tag{3.20}$$

here β_l is a given constant. Recall that $\rho_s^*[u]\equiv\omega_d^0(\varepsilon_s k_s^2\frac{\Delta\varphi_\infty}{2}sg-q_s[u])$.

For equation (3.20) we shall establish existence of a unique continuous and bounded solution, via the contraction mapping argument. Let us introduce the notations $Q_d^s[\varphi](x)=((\omega_d^0 q_s[\varphi])*\psi_s^0)(x)$ and

$$f_s(x)=\Delta\varphi_\infty\left(\frac{sg(x)}{2}[1-\exp(-k_s|x|)]+\psi_s^*(x)-\frac{\varepsilon_s k_s^2}{2}((\omega_d^0.sg)*\psi_s^0)(x)+W_L\,[\psi_s^{0,1}](x)\right)+\beta_l\psi_s^0(x)\,. \tag{3.21}$$

Now, substituting $\psi_s^{0,1}$ in (3.20) with the right hand side of (3.15), equation (3.20) takes the form

$$\varphi-\varphi_\infty^*-Q_d^s[\varphi]=f_s\,. \tag{3.22}$$

To analyze the above equation, we shall use the norm $||w||\equiv\sup_x|w(x)|$, $x\in R^1$, for continuous, bounded functions $w(x)$ on R^1, and shall deal with balls $B_r(\varphi_\infty^*)$, centered at

φ_∞^*, having radius r; here $B_r(\varphi_\infty^*)$ is the closure, regarding the norm $||.||$, to the set of the bounded continuous functions $v(x), x \in R^1$, such that $||v - \varphi_\infty^*|| \leq r$, for a fixed r. It is clear that under the norm $||.||$ $B_r(\varphi_\infty^*)$ is a complete metric space. We have to study the map $\varphi \to \Phi$, $\Phi = \Phi[\varphi] \equiv \varphi_\infty^* + Q_d^s[\varphi] + f_s$, for $\varphi \in B_r(\varphi_\infty^*)$. For choosing a proper magnitude of radius r, we shall take into account that $Q_d^s[\varphi_\infty^*] = 0$ and $f_s = \psi_s^0$ in case of the linear canonical form to problem (3.4), (3.5) (possessing ψ_s^0 as the unique solution). Thus, observing that $||\psi_s^0|| = \psi_s^0(0) \leq \dfrac{1}{2\varepsilon_s k_s}$, we shall fix below a final choice of r in the form $\dfrac{1+\Delta r}{2\varepsilon_s k_s}$ (Δr - a small positive parameter). We begin the estimation to the image deviation $\Phi - \varphi_\infty^*$, $\Phi = \Phi[\varphi]$, by the obvious triangle inequality:

$$||\Phi - \varphi_\infty^*|| \leq ||Q_d^s[\varphi]|| + ||f_s||, \tag{3.23}$$

for $\varphi \in B_r(\varphi_\infty^*)$. By the inverse Fourier transformation of $\hat{\varphi}_Q(\xi)$ ($\varphi_Q = Q_d^s[\varphi]$) we have $|Q_d^s[\varphi](x)| \leq ||q_s[\varphi]||\,.\psi_s^0(0)$ (via the given definition of $Q_d^s[\varphi]$, formula (3.13) – for ψ_s^0 and equality $\int_{-\infty}^{+\infty} \omega_d^0(x)dx = 1$), consequently it holds: $||Q_d^s[\varphi]|| \leq ||q_s[\varphi]|| \dfrac{1}{2\varepsilon_s k_s}$. Next, from $||\varphi - \varphi_\infty^*|| \leq r$ we find for $q_s[\varphi]$ that $||q_s[\varphi]|| \leq \dfrac{(\varepsilon_s k_s^2 r)^2}{2}\left(1 + \dfrac{\varepsilon_s k_s^2 r}{3}\right)$. Choosing now $\Delta r = \dfrac{1}{2}$, we can fix the magnitude of r:

$$r = \frac{3}{4\varepsilon_s k_s}. \tag{3.24}$$

Then we find the next inequality, for $||q_s[\varphi]||$:

$$||q_s[\varphi]|| \leq \frac{9k_s^2}{32}\left(1 + \frac{3k_s}{8}\right). \tag{3.25}$$

Under condition $\dfrac{3}{4} \leq k_s^{-1}$ (applied below as a contraction requirement to $\Phi[\varphi]$) inequality (3.25) can be easily reworked till a convenient estimate for $||q_s[\varphi]||$. We shall introduce, for a sake of simplicity, also a restriction in the form $k_s^{-1} \leq \kappa'$ (with an arbitrary constant $\kappa' > \dfrac{3}{4}$). For a large class of semiconductors (including these in [5], [7]) it is enough to take $\kappa' = 2$. Thus we can suppose from now on that

$$\frac{3}{4} \leq k_s^{-1} \leq 2. \tag{3.26}$$

Now, reworking (3.25) we get: $||q_s[\varphi]|| \le \frac{3}{4}$. Consequently $Q_d^s[\varphi]$ is estimated as

$$||Q_d^s[\varphi]|| \le \frac{r}{2}. \tag{3.27}$$

The obvious next step is to establish the analogous estimate for f_s, to get this way the needed property for $\Phi[\varphi]$ (see (3.23)). For relevant terms with $\Delta\varphi_\infty$ in expression (3.21) it holds as follows. Function $\psi_s^*(x)$ satisfies (as directly shows (3.13)) inequality $||\psi_s^*|| \le \psi_s^0(0)$, therefore $||\psi_s^*|| \le \frac{1}{2\varepsilon_s k_s}$. Next, estimating the term $(\omega_d^0.sg)*\psi_s^0$ by analogy to $Q_d^s[\varphi]$, we have:

$$||(\omega_d^0.sg)*\psi_s^0|| \le \frac{1}{2\pi}\int_{-\infty}^{+\infty}\omega_d^0(x)dx\int_{-\infty}^{+\infty}|\hat{\psi}_s^0(\xi)|\,d\xi \le \frac{1}{2\varepsilon_s k_s}.$$

Concerning the element $W_L[\psi_s^{0,1}]$ (in said expression for f_s) we shall firstly introduce inequality

$$||W_L[\psi_s^{0,1}]|| \le \frac{1}{2\pi}||2|\xi|(k_s^2+\xi^2)^{-1}||_{L_2}||\hat{\psi}_s^{0,1}||_{L_2}$$

(where $||.||_{L_2}$ is the norm in $L_2(R^1)$). Then we use that $||2|\xi|(k_s^2+\xi^2)^{-1}||_{L_2} = \left(\frac{2\pi}{k_s}\right)^{1/2}$ and, because of (3.15), we have $||\hat{\psi}_s^{0,1}||_{L_2} \le ||\hat{\psi}_s^{0,1} - \frac{1}{\varepsilon_s}(W_L[\psi_s^{0,1}]\hat{)}||_{L_2} \le \frac{1}{2\varepsilon_s}||(sg.\exp(-k_s|.|\hat{)}||_{L_2}$ (with $(\Phi\hat{)}$ as the Fourier transformation of Φ). The above yields inequality $||W_L[\psi_s^{0,1}]|| \le \frac{1}{\varepsilon_s k_s\sqrt{2}} < \frac{3}{4\varepsilon_s k_s}$, taking into account that $||(sg.\exp(-k_s|.|\hat{)}||_{L_2} = 2\left(\frac{\pi}{k_s}\right)^{1/2}$; i.e. $||W_L[\psi_s^{0,1}]|| \le r$ (see (3.24)). Then, from inequality (see (3.21))

$$||f_s|| \le |\Delta\varphi_\infty|\left(\frac{1}{2} + ||\psi_s^*|| + \frac{\varepsilon_s k_s^2}{2}||(\omega_d^0.sg)*\psi_s^0|| + ||W_L[\psi_s^{0,1}]||\right) + |\beta_l|\,||\psi_s^0||,$$

it follows the next one: $||f_s|| \le \left(\frac{1}{2} + \frac{2}{3}r + \frac{\varepsilon_s k_s^2}{4\varepsilon_s k_s} + r\right)|\Delta\varphi_\infty| + \frac{2r}{3}|\beta_l|$. Consequently, at $2r \ge 1$, i.e. (via (3.24)) introducing condition (3.28), below, we find firstly that $||f_s|| \le r\left([\frac{8}{3} + \frac{\varepsilon_s k_s^2}{3}]|\Delta\varphi_\infty| + \frac{2}{3}|\beta_l|\right)$.

$$\varepsilon_s k_s \leq \frac{3}{2}. \tag{3.28}$$

By (3.28) the above found for $||f_s||$ modifies to inequality $||f_s|| \leq \frac{r}{3}(14|\Delta\varphi_\infty|+2|\beta_l|)$. This will give the estimate

$$||f_s|| \leq \frac{r}{3}. \tag{3.29}$$

It holds when the sum $7|\Delta\varphi_\infty|+|\beta_l|$ satisfies condition

$$7|\Delta\varphi_\infty|+|\beta_l| \leq \frac{1}{2}. \tag{3.30}$$

Summarizing (3.23), (3.27) and (3.29), we conclude the following. For any data φ_∞^+, φ_∞^-, β_l and parameters k_s, ε_s satisfying (3.26), (3.28) and (3. 30) relation $\Phi = \Phi[\varphi]$ maps the ball $B_r(\varphi_\infty^*)$, with r as in (3.24), into itself.

Now we shall study the contraction property of $\Phi[\varphi]$. For arbitrary two elements $\varphi_1, \varphi_2 \in B_r(\varphi_\infty^*)$ we have to estimate difference $\Phi_2 - \Phi_1$, $\Phi_j = \Phi[\varphi_j]$ $(j=1,2)$. From $\Phi_2 - \Phi_1 = Q_d^s[\varphi_2] - Q_d^s[\varphi_1]$ we shall consider difference of $Q_d^s[\varphi_j]$ $(j=1,2)$, using relation

$$Q_d^s[\varphi_2] - Q_d^s[\varphi_1] = \left(\omega_d^0(\varphi_2-\varphi_1)\int_0^1 q_s'[\varphi_1+\tau(\varphi_2-\varphi_1)]d\tau\right) * \psi_s^0. \tag{3.31}$$

(Above $q_s'[t]$ is derivative $\frac{dq_s}{dt}[t]$, with $q_s[t] \equiv \frac{1}{2!}[\varepsilon_s k_s^2(t-\varphi_\infty^*)]^2 - \frac{1}{3!}[\varepsilon_s k_s^2(t-\varphi_\infty^*)]^3$.) Denote by $R_d^s[\varphi_{2,1}] * \psi_s^0$ the right hand side of (3.31) and take into account the next several inequalities:

$$||Q_d^s[\varphi_2] - Q_d^s[\varphi_1]|| \leq \frac{1}{2\pi}\int_{-\infty}^{+\infty} |\hat{R}_d^s[\varphi_{2,1}](\xi)|.|\hat{\psi}_s^0(\xi)|\,d\xi\,;$$

$$||\hat{R}_d^s[\varphi_{2,1}]|| \leq ||\varphi_2-\varphi_1|| \int_{-\infty}^{+\infty} \omega_d^0(x)\int_0^1 |q_s'[\varphi_1+\tau(\varphi_2-\varphi_1)](x)|\,d\tau dx\,;$$

$$||q_s'[\cdot]|| \leq \varepsilon_s k_s^2 r\left(1+\frac{\varepsilon_s k_s^2 r}{2}\right) \text{ (at } \varphi_1, \varphi_2 \in B_r(\varphi_\infty^*)\text{), i.e. } ||q_s'[\cdot]|| \leq \frac{3k_s}{4}\left(1+\frac{3k_s}{8}\right) \leq \frac{3}{2};$$

$$\int_0^1 ||q_s'[\varphi_1+\tau(\varphi_2-\varphi_1)]||\,d\tau \leq \frac{3}{2}. \tag{3.32}$$

(We have used in (3.32) relation (3.26), via (3.24), and the above notations $\hat{\psi}_s^0$, $\hat{R}_d^s[\varphi_{2,1}]$ are taken for the Fourier images respectively to functions $\psi_s^0(x)$, $R_d^s[\varphi_{2,1}](x)$.) Then we have:

$$||Q_d^s[\varphi_2]-Q_d^s[\varphi_1]|| \le \frac{3}{2}||\varphi_2-\varphi_1||\frac{1}{2\pi}\int_{-\infty}^{+\infty}|\hat{\psi}_s^0(\xi)|d\xi \le \frac{3}{4\varepsilon_s k_s}||\varphi_2-\varphi_1||.$$

Consequently,

$$||\Phi_2-\Phi_1|| \le r||\varphi_2-\varphi_1||$$

(because of (3.24)), and under estimates (3.27), (3.29) (valid at (3.30)) the considered map $\Phi = \Phi[\varphi]$ is a contraction in the ball $B_r(\varphi_\infty^*)$, for $r<1$, i.e. $\frac{3}{4}<\varepsilon_s k_s$. Combining the latter inequality with (3.28), we come to condition

$$\frac{3}{4}< \varepsilon_s k_s \le \frac{3}{2}. \tag{3.33}$$

Thus we have given the proof of the following basic result.

3.2 Proposition

For arbitrary values of positive parameters ε_s, k_s, each asymptotic data $\varphi_\infty^\pm$ of the surface potential and line charges $\rho_l = \varepsilon_0\beta_l$, such that conditions (3.26), (3.33) and (3.30) hold, equation (3.20) possesses a unique continuous, bounded solution $\varphi(x;d)$, satisfying the estimate

$$\sup_x|\varphi(x;d)-\varphi_\infty^*| \le \frac{5}{8\varepsilon_s k_s}, \forall d>0. \tag{3.34}$$

4. Explicit approximations in the case of semiconductor interface

Via the possible applications, it is important to ask for a suitable approximation $\varphi_*^0(x)$ to the interface data $u(x,0)$, well enough at small $|d|$ and explicitly determined. To that goal, suppose a sequence $\{\varphi_n\}$ of solutions to (3.20), $\varphi_n(x)\equiv\varphi(x;d)$, at $d=d_n$, with $d_n\to 0$ ($n\to\infty$), is convergent (in a distribution sense) to a bounded continuous function $\varphi_*(x)$, $x\in R^1$. Putting $d=d_n$ in (3.20) and letting $n\to\infty$, we can conclude that $\varphi_*(x)$ is a solution to equation

$$\varphi(x)=\varphi_\infty+(\beta_l+q_s[\varphi(0)])\psi_s^0(x)+\Delta\varphi_\infty(\psi_s^*(x)+\varepsilon_s\psi_s^{0,1}(x)). \tag{4.1}$$

For finding (4.1) we have taken into account that $Q_d^s[\varphi(.;d)](x)\to q_s[\varphi_*(0)]\psi_s^0(x)$ and $[(\omega_d^0.sg)*\psi_s^0](x)\to 0$, $\forall x\in R^1$, at $d=d_n$ ($n\to\infty$). Next we shall study equation (4.1). Note first of all the necessary condition to have a continuous solution $\varphi(x)$:

$$\varphi_\infty^- + \varepsilon_s\Delta\varphi_\infty\psi_s^{0,1}(-0)=\varphi_\infty^+ + \varepsilon_s\Delta\varphi_\infty\psi_s^{0,1}(+0);$$

it is fulfilled, because of (3.15). If $\varphi(x)$ is a continuous solution to (4.1), for value $\varphi(0)$ we obtain (from (4.1), at $x=+0$ or $x=-0$) the algebraic equation

$$\varphi(0) = \varphi_\infty^* + (\beta_l + q_s[\varphi(0)])\psi_s^0(0)\,;$$

i.e. $\varphi(0)$ is a real solution to equation

$$\psi_s^0(0)\frac{(\varepsilon_s k_s^2)^3}{6}(z-\varphi_\infty^*)^3 - \psi_s^0(0)\frac{(\varepsilon_s k_s^2)^2}{2}(z-\varphi_\infty^*)^2 + z - \varphi_\infty^* - \beta_l\psi_s^0(0) = 0\,.$$

Setting $t = \varepsilon_s k_s^2(z-\varphi_\infty^*)$ we rewrite this equation in the form:

$$\frac{t^3}{6} - \frac{t^2}{2} + \frac{t}{\varepsilon_s k_s^2 \psi_s^0(0)} - \beta_l = 0\,. \tag{4.2}$$

Derivative of the left hand side (denote it by $g(t)$) is $g'(t) = \frac{t^2}{2} - t + \frac{1}{\varepsilon_s k_s^2 \psi_s^0(0)}$. The found quadratic polynomial does not have real roots (via (3.26) and the known inequality $\psi_s^0(0) \le \frac{1}{2\varepsilon_s k_s}$, recall (3.13) concerning $\psi_s^0(0)$). This yields existence of a unique real solution t^0 of (4.2) and we set

$$\varphi^0 = \varphi_\infty^* + \frac{t^0}{\varepsilon_s k_s^2}\,. \tag{4.3}$$

Now from (4.1) we get the function

$$\varphi_*^0(x) \equiv \varphi_\infty + (\beta_l + q_s[\varphi^0])\psi_s^0(x) + \Delta\varphi_\infty(\psi_s^*(x) + \varepsilon_s \psi_s^{0,1}(x))\,. \tag{4.4}$$

It presents actually the unique solution of equation (4.1).

The next step will be the comparison of $\varphi_*^0(x)$ and $\varphi(x;d)$. Let us introduce the difference $\Delta Q_d^s[\varphi](x) \equiv Q_d^s[\varphi](x) - q_s[\varphi(0)]\psi_s^0(x)$. Formula (4.4) then directly shows that $\varphi_*^0(x)$ is a solution to equation

$$\varphi(x) - \varphi_\infty^* - Q_d^s[\varphi](x) = f_s(x) - \Delta Q_d^s[\varphi](x)\,. \tag{4.5}$$

Subtracting (4.5), with $\varphi = \varphi_*^0$, from (3.20) – with $\varphi = \varphi(.;d)$, we evidently get:

$$\varphi(.;d) - \varphi_*^0 = Q_d^s[\varphi(.;d)] - Q_d^s[\varphi_*^0] + \Delta Q_d^s[\varphi_*^0]\,.$$

Putting afterwards $\varphi_2 = \varphi(.;d)$ and $\varphi_1 = \varphi_*^0$ in the above given contraction estimate – for $Q_d^s[\varphi_2] - Q_d^s[\varphi_1]$, we find directly that

$$||\varphi(.;d) - \varphi_*^0|| \le r||\varphi(.;d) - \varphi_*^0|| + ||\Delta Q_d^s[\varphi_*^0]||,$$

consequently

$$||\varphi(.;d) - \varphi_*^0|| \le \frac{1}{1-r}||\Delta Q_d^s[\varphi_*^0]||\,. \tag{4.6}$$

By the known definition $Q_d^s[\varphi] \equiv (\omega_d^0 q_s[\varphi]) * \psi_s^0$, perturbation term $\Delta Q_d^s[\varphi_*^0]$ can be easily presented as

$$\begin{aligned}\Delta Q_d^s[\varphi_*^0](x) = \int_{-\infty}^{+\infty} \omega^0(\tau)(q_s[\varphi_*^0(\tau d)] - q_s[\varphi^0])\psi_s^0(x-\tau d)d\tau + \\ +q_s[\varphi^0]\int_{-\infty}^{+\infty} \omega^0(\tau)[\psi_s^0(x-\tau d) - \psi_s^0(x)]d\tau\end{aligned} \quad . \tag{4.7}$$

Denote the first and second integrals in (4.7) respectively by $I_1[\varphi_*^0](x)$ and $I_2[\varphi_*^0](x)$; they satisfy the following inequalities:

$$||I_1[\varphi_*^0]|| \le 2||\psi_s^0|| \sup_{|y|\le d} |q_s[\varphi_*^0(y)] - q_s[\varphi^0]|; \tag{4.8.a}$$

$$||I_2[\varphi_*^0]|| \le |q_s[\varphi^0]| \int_{-\infty}^{+\infty} \omega^0(\tau) \sup_x |\psi_s^0(x-\tau d) - \psi_s^0(x)| d\tau . \tag{4.8.b}$$

To rework estimate (4.8.a), taking the arguments, known from the analysis of (3.29) (see above), we use at the beginning that

$$||q_s[\varphi_*^0] - q_s[\varphi^0]||_d \le \varepsilon_s k_s^2 (1 + \frac{\varepsilon_s k_s^2}{2}) ||\varphi_*^0 - \varphi^0||_d ,$$

where

$$||\varphi_*^0 - \varphi^0||_d \equiv \sup_{|y|\le d} |\varphi_*^0(y) - \varphi^0| ; \text{ i.e.}$$

$$||q_s[\varphi_*^0] - q_s[\varphi^0]||_d \le \frac{3k_s}{2}(1 + \frac{3}{4}k_s) ||\varphi_*^0 - \varphi^0||_d \le 4||\varphi_*^0 - \varphi^0||_d ,$$

via (3.26), (3.33). Substituting then in (4.8.a), we have:

$$||I_1[\varphi_*^0]|| \le 8||\psi_s^0||.||\varphi_*^0 - \varphi^0||_d . \tag{4.9.a}$$

(We shall show afterwards that $||\varphi_*^0 - \varphi^0||_d \le 1$ at small enough d.) Next, for reworking of (4.8.b), we shall firstly apply equality

$$\psi_s^0(x-y) - \psi_s^0(x) = -y\int_0^1 \psi_s^{0,1}(x-ty)dt .$$

(Via formula (3.13) for $\psi_s^0(x)$ it is not difficult to verify validity of the above.) Consequently $\sup_x |\psi_s^0(x-y) - \psi_s^0(x)| \le |y|.||\psi_s^{0,1}||$, and, because of (4.8.b), we find inequality

$$||I_2[\varphi_*^0]|| \le |q_s[\varphi^0]|.||\psi_s^{0,1}|| d\int_{-\infty}^{+\infty} |\tau| \omega^0(\tau)d\tau ,$$

which actually yields that:

$$||I_2[\varphi_*^0]|| \leq |q_s[\varphi^0]|.||\psi_s^{0,1}||d. \tag{4.9.b}$$

To final reworking of (4.8.a) we shall estimate the quantity $||\varphi_*^0 - \varphi^0||_d$ in (4.9.a). By (4.4) and the algebraic equation $\varphi(0) = \varphi_\infty^* + (\beta_l + q_s[\varphi(0)])\psi_s^0(0)$ for $\varphi(0)$ (see the initial form of (4.2)) we firstly have

$$\varphi_*^0(x) - \varphi^0 = \frac{\Delta\varphi_\infty}{2} sg(x) + (\beta_l + q_s[\varphi^0])[\psi_s^0(x) - \psi_s^0(0)] + \Delta\varphi_\infty[\psi_s^*(x) + \varepsilon_s\psi_s^{0,1}(x)],$$

and expressing $\psi_s^{0,1}$ by (3.15) we get

$$\varphi_*^0(x) - \varphi^0 = \frac{\Delta\varphi_\infty}{2} sg(x)[1 - \exp(k_s|x|)] + (\beta_l + q_s[\varphi^0])[\psi_s^0(x) - \psi_s^0(0)] + \tag{4.10}$$

$$+\Delta\varphi_\infty(\psi_s^*(x) + W_L[\psi_s^{0,1}](x)).$$

Next we analyze the terms in (4.10) beginning with function $W_L[\psi_s^{0,1}](x)$; it is continuous and odd, and we can use the following inequalities:

$$|W_L[\psi_s^{0,1}]|(x) \leq \frac{2}{\pi}\int_0^{+\infty} \frac{\xi|\sin(x\xi)|}{k_s^2 + \xi^2}|\hat{\psi}_s^{0,1}(\xi)|\,d\xi \leq \frac{1}{\pi}\left(\int_0^{+\infty} \frac{\xi^2|\sin^2(x\xi)|}{(k_s^2 + \xi^2)^2}d\xi\right)^{1/2}.||\hat{\psi}_s^{0,1}||_{L_2(R^1)}.$$

Therefore:

$$|W_L[\psi_s^{0,1}]|(x) \leq \frac{\sqrt{|x|}}{\pi}\left(\int_0^{+\infty} \frac{\sin^2\theta}{\theta^2}d\theta\right)^{1/2}.||\hat{\psi}_s^{0,1}||_{L_2(R^1)}.$$

Recall here the already found estimate $||\hat{\psi}_s^{0,1}||_{L_2(R^1)} \leq \dfrac{\sqrt{\pi}}{\varepsilon_s\sqrt{k_s}}$ and the well known relation $\int_0^{+\infty} \dfrac{\sin^2\theta}{\theta^2}d\theta = \dfrac{\pi}{2}$. Then it follows:

$$||W_L[\psi_s^{0,1}]||_d \leq \frac{\sqrt{d}}{\varepsilon_s\sqrt{k_s}}. \tag{4.11}$$

For $\psi_s^*(x)$ we start with the inequality

$$|\psi_s^*(x)| \leq \frac{|x|}{\pi\varepsilon_s}\int_0^{1/d}\left|\frac{\sin(x\xi)}{x\xi}\right|\frac{\xi}{k_s^2 + \xi^2}d\xi + \frac{1}{\pi\varepsilon_s}\int_{1/d}^{\infty}\frac{|\sin(x\xi)|}{k_s^2 + \xi^2}d\xi;$$

i.e.:

$$|\psi_s^*(x)| \leq \frac{|x|}{2\pi\varepsilon_s}\log\left(1 + \frac{1}{k_s^2 d^2}\right) + \frac{1}{\pi\varepsilon_s}\int_{1/d}^{\infty}\frac{1}{\xi^2}d\xi \leq \frac{|x|}{\pi\varepsilon_s}\log\left(1 + \frac{1}{k_s d}\right) + \frac{d}{\pi\varepsilon_s}.$$

Evidently, $\log(1+\frac{1}{k_s d}) \leq \frac{1}{\sqrt{k_s d}}$, and we find:

$$||\psi_s^*||_d \leq \frac{1}{\pi\varepsilon_s}(d+\frac{\sqrt{d}}{\sqrt{k_s}}). \tag{4.12}$$

The estimate for $\psi_s^0(x)-\psi_s^0(0)$ is a consequence from the above one for $\psi_s^0(x-y)-\psi_s^0(x)$, thus we come to inequality

$$||\psi_s^0-\psi_s^0(0)||_d \leq d||\psi_s^{0,1}||, \tag{4.13}$$

and for $1-\exp(k_s|x|)$ we can take the next obviously one:

$$||1-\exp(k_s|x|)||_d \leq k_s d\exp(k_s d). \tag{4.14}$$

Now (4.10) and (4.11) - (4.14) yield:

$$||\varphi_*^0-\varphi^0||_d \leq |\Delta\varphi_\infty|(\frac{\sqrt{k_s d}}{2}\exp(k_s d)+\frac{4d\sqrt{d}}{9\varepsilon_s^2\sqrt{k_s}})+(|\beta_l|+|q_s[\varphi^0]|)||\psi_s^{0,1}||d.$$

The above simplifies, at $dk_s \leq 1$, to inequality

$$||\varphi_*^0-\varphi^0||_d \leq |\Delta\varphi_\infty|(\frac{3}{2}+\frac{20}{9}r)\sqrt{dk_s}+(|\beta_l|+|q_s[\varphi^0]|)\frac{3d}{4\varepsilon_s}(1+\frac{1}{\varepsilon_s k_s});$$

i.e.

$$||\varphi_*^0-\varphi^0||_d \leq [6|\Delta\varphi_\infty|+(1+\frac{4}{3})(|\beta_l|+|q_s[\varphi^0]|)]r\sqrt{dk_s},$$

and

$$||\varphi_*^0-\varphi^0||_d \leq 3(2|\Delta\varphi_\infty|+|\beta_l|+|q_s[\varphi^0]|)\sqrt{dk_s}. \tag{4.15}$$

Going to expression (4.7), we have firstly $||\Delta Q_d^s[\varphi_*^0]|| \leq ||I_1[\varphi_*^0]||+||I_2[\varphi_*^0]||$ and applying afterwards (4.9.a), combined with (4.15) and (4.9.b), we establish the estimate:

$$||\Delta Q_d^s[\varphi_*^0]|| \leq 8||\psi_s^0||.||\varphi_*^0-\varphi^0||_d+|q_s[\varphi^0]|.||\psi_s^{0,1}||d. \tag{4.16}$$

Next, for the relevant quantities in (4.16) it can be easily established (via (3.26), (3.33)) as follows:

$$8||\psi_s^0||.||\varphi_*^0-\varphi^0||_d \leq \frac{12}{\varepsilon_s k_s}(2|\Delta\varphi_\infty|+|\beta_l|+|q_s[\varphi^0]|)\sqrt{dk_s} \leq 16(2|\Delta\varphi_\infty|+|\beta_l|+|q_s[\varphi^0]|)\sqrt{dk_s};$$

$$|q_s[\varphi^0]|.||\psi_s^{0,1}||d \leq (1+\frac{4}{3})|q_s[\varphi^0]|\frac{3d}{4\varepsilon_s} \leq 3|q_s[\varphi^0]|\sqrt{dk_s}.$$

Applying the above two inequalities to (4.16), we find that

$$||\Delta Q_d^s[\varphi_*^0]|| \le 8(4|\Delta\varphi_\infty| + 2|\beta_l| + 3|q_s[\varphi^0]|)\sqrt{dk_s}\,. \tag{4.17}$$

Finally, from (4.6), (4.17) we directly get:

$$||\varphi(.;d) - \varphi_*^0|| \le \frac{8}{1-r}(4|\Delta\varphi_\infty| + 3|q_s[\varphi^0]| + 2|\beta_l|)\sqrt{dk_s}\,.$$

Now let us introduce (for a simplicity sake) the restriction $0.5 \le r \le 0.9$, equivalent to condition

$$\frac{5}{6} \le \varepsilon_s k_s \le \frac{3}{2}\,. \tag{4.18}$$

Then $1-r \ge 0.1$, consequently $\frac{8}{1-r} \le 80$, and, from the above inequality for $\varphi(.;d) - \varphi_*^0$, we obtain the needed estimate, approximating for the exact solution $\varphi(x;d)$,at $d \to 0$:

$$||\varphi(.;d) - \varphi_*^0|| \le 80(2|\beta_l| + 3|q_s[\varphi^0]| + 4|\Delta\varphi_\infty|)\sqrt{dk_s}\,. \tag{4.19}$$

By the above arguments we have actually proven the following assertion:

4.1 Proposition

Function $\varphi_*^0(x)$ is an approximation to solution $\varphi(x;d)$ of equation (3.20), at $d \to 0$, explicitly determined by formula (4.4) and satisfying estimate (4.19), for parameters $\varphi_\infty^\pm, \beta_l$, which fulfill (3.30) and (ε_s, k_s) varying in the compact set determined by (3.26), (4.18).

5. Concluding remarks

Here we accent on the approximating solutions, in several applicable variants, via the convenience of solution determination by effective formulas (see positions 2) – 5), below). Note, as a principle, that the possible explicit solutions (presenting for instance the interface electric potential) are necessary for examination of relevant numerical methods, and the same holds for the explicit approximations to the exact implicit solutions. Below we start with some dimensional remarks, related in particular to known experimental data.

1. In a $(k_s^{-1}, \varepsilon_s)$ - coordinate system, scaled in nanometers, the above mentioned compact is trapezoid, with contours – the straight lines: $k_s^{-1} = \frac{3}{4}nm$ and $k_s^{-1} = 2\,nm$ (as the bottoms of trapezoid, vertically situated), and $\varepsilon_s = \frac{5}{6}k_s^{-1}nm$, $\varepsilon_s = \frac{3}{2}k_s^{-1}nm$ (as the thighs). For the classes of semiconductors analyzed in [5], [7] the values of parameter k_s^{-1} are not greater than 1 nm, satisfying thus condition (3.26) in the form

$$0.75\,nm \le k_s^{-1} \le 2\,nm\,. \tag{4.20}$$

The key non-dimensional quantity in the surface electrostatics is given by the product $\varepsilon_s k_s$, and the same holds for the above used dk_s. On the other hand, quantity $(\varepsilon_s k_s)^{-1}$ can be automatically provided with a (preferable) voltage dimension (see also expression (3.24) of r). Recall that such a mechanism has been suggested by the estimate $\|\psi_s^0\| = \psi_s^0(0) \le \frac{1}{2\varepsilon_s k_s}$ of canonical surface potential ψ_s^0. This allows, for mathematical reworks, to use the product $k_s r$ (in the important factor $\varepsilon_s k_s^2 r = \varepsilon_s k_s . k_s r$) as non-dimensional.

2. The proposed model (2.2) - (2.6), with $\varepsilon_b^- = 1, k_b^- = 0$, admits explicit approximations $\varphi_*^0(x)$ and $u_*^0(x,z)$,

$$\varphi(x;d) = \varphi_*^0(x) + O(\sqrt{dk_s})\,(d \to 0),\ \forall x \in R^1;$$

$$u(x,z) = u_*^0(x,z) + O(\sqrt{dk_s})\,(d \to 0),\ \forall (x,z) \in R^2.$$

They satisfy estimates (4.19) and

$$\sup_z \| u(.,z) - u_*^0(.,z) \| \le 80(2|\beta_l| + 3|q_s[\varphi^0]| + 4|\Delta\varphi_\infty|)\sqrt{dk_s}\,. \tag{4.21}$$

In addition, said approximations are determined respectively by formulas (4.4) and

$$u_*^0(x,z) = \begin{cases} \frac{|z|}{\pi} \int_{-\infty}^{+\infty} \frac{\varphi_*^0(t)}{(x-t)^2 + z^2} dt, & z \ne 0; \\ \varphi_*^0(x), & z = 0 \end{cases}. \tag{4.22}$$

3. In case of relatively small $\Delta\varphi_\infty$, i.e. $|\Delta\varphi_\infty| << |\varphi_\infty^*|$, the term $\Delta\varphi_\infty(\psi_s^*(x) + \varepsilon_s \psi_s^{0,1}(x))$ can be neglected in representation (4.4) and we can use the simplified approximation $\varphi_{0,*}(x) \equiv \varphi_\infty + (\beta_l + q_s[\varphi^0])\psi_s^0(x)$ of $\varphi(x;d)$, instead of $\varphi_*^0(x)$. This yields the simpler approximation $u_{0,*}(x,z)$ to the space potential $u(x,z)$, with

$$u_{0,*}(x,z) = \varphi_\infty(x) + \frac{\beta_{l,s}}{\pi} \int_0^{+\infty} \frac{\exp(-|z|\xi)\cos(x\xi)}{2\xi + \varepsilon_s(k_s^2 + \xi^2)} d\xi,\ z \in R^1. \tag{4.23}$$

Above $\beta_{l,s} = \beta_l + q_s[\varphi^0]$. At $z = 0$ formula (4.23) evidently gives $u_{0,*}(x,0) = \varphi_{0,*}(x)$. Here it should be specially noted that known real situations (see for instance in [5], [7]) are contained in the case $\Delta\varphi_\infty = 0$.

4. The case $\varphi_\infty^- = \varphi_\infty^+ = 0$ (then $\Delta\varphi_\infty = 0$ and $\varphi_\infty^* = 0$) covers the experimental models in [5], [7]. Now it seems to be an open question whether the line phase charges can get essentially smaller values than these for the surface zones (called terraces) $\{z = 0, d << |x|\}$ - after specific annealing of indium-phosphorus semiconductor

samples, say ([5], [7]). Then we would have relatively small values of $|\beta_l|$, terms with t^2 and t^3 (at $t = t^0$) can be neglected in (4.2), and we can take the value of $p_{s,0}\beta_l$ (with $p_{s,0} = \psi_s^0(0)$) as an approximation of $\dfrac{t^0}{\varepsilon_s k_s^2}$ (using (4.2)). It gives that $q_s[\varphi^0] \approx \frac{1}{2}\varepsilon_s^2 k_s^4 p_{s,0}^2 \beta_l^2$. Replace in (4.23) $\beta_{l,s}$ with $\beta_l + \frac{1}{2}\varepsilon_s^2 k_s^4 p_{s,0}^2 \beta_l^2$, and take $\psi_s^0(0) \approx \dfrac{1}{2\varepsilon_s k_s}$ (via the known estimate $\psi_s^0(0) \le \dfrac{1}{2\varepsilon_s k_s}$). Thus we can consider the function $w_{0,*}(x,z)$, below, as a next approximation of the exact potential $u(x,z)$.

$$w_{0,*}(x,z) = \frac{\beta_l + 8^{-1}.k_s^2\beta_l^2}{\pi}\int_0^{+\infty}\frac{\exp(-|z|\xi)\cos(x\xi)}{2\xi + \varepsilon_s(k_s^2 + \xi^2)}d\xi,\ z \in R^1. \tag{4.24}$$

The found formula conveniently shows that the (nonlinear) impact of the vacancy denuded sub-strips $\{z = 0, \theta d < |x| < d\}$ (with $0 < \theta < 1$) is compatible to the perturbation $\dfrac{k_s^2\beta_l^2}{8\pi}\displaystyle\int_0^{+\infty}\frac{\exp(-|z|\xi)\cos(x\xi)}{2\xi + \varepsilon_s(k_s^2 + \xi^2)}d\xi,\ z \in R^1$.

5. A special variant is presented by the case of weakly charged contour $\{z = 0, x = 0\}$, combined with a relatively higher asymptotic surface power $\Delta\varphi_\infty$. From experimental view point (c.f. [5], [7]) said situation seems to be another open question. Now we can assume that $|\beta_l| << |\Delta\varphi_\infty|$. Then neglecting term $(\beta_l + q_s[\varphi^0])\psi_s^0(x)$ in the expression for $\varphi_*^0(x)$, we insert in (4.22) $\varphi_\infty(t) + \Delta\varphi_\infty(\psi_s^*(t) + \varepsilon_s\psi_s^{0,1}(t))$ (instead of $\varphi_*^0(t)$) and find the following expression:

$$w_{0,\infty}(x,z) = \varphi_\infty(x) - \frac{\Delta\varphi_\infty}{\pi}\int_0^{+\infty}\frac{\exp(-|z|\xi)(1+\varepsilon_s\xi)\sin(x\xi)}{2\xi + \varepsilon_s(k_s^2 + \xi^2)}d\xi,\ z \ne 0. \tag{4.26}$$

Here $w_{0,\infty}(x,0) = \varphi_\infty(x) + \Delta\varphi_\infty(\psi_s^*(x) + \varepsilon_s\psi_s^{0,1}(x))$. Modified potential $w_{0,\infty}(x,z)$ presents the impact of asymptotic power $\Delta\varphi_\infty$ on the space potential distribution.

6. Acknowledgement

The important extension of the Bedeaux – Vlieger formal scheme into the larger one – for decomposing of different dimensional singularities, is due to Prof. B. Radoev (University of Sofia, Bulgaria, Dept. of Physical Chemistry). The author is grateful to Dr. Plamen Georgiev and Dr. Emil Molle (University of Sofia, Bulgaria, Faculty of Biology) for the useful comments on cell biology concepts and the assistance in preparing the illustrations.

This study was partially supported by grant No DDVU 02/90 of the Bulgarian National Science Foundation.

7. References

[1] Ashcroft, N.W., Mermin, N.D., Solid States Physics, Holt, Rinehart and Winston - New York (1975).

[2] Bedeaux, D., Vlieger, J., Optical Properties of Surfaces, Imperial College Press, London(2001).

[3] Colton, D., Kress, R., Integral Equation Methods in Scattering Theory, John Wiley & Sons, New York (1983).

[4] Cook, B., Kazakova, T., Madrid, P., Neal, J., Pauletti, M., Zhao, R., Cell-foreign Particle Interaction, IMA Preprint Series # 2133-3 (Sept. 2006), Univ. of Minnesota.

[5] Ebert, Ph., Hun Chen, Heinrich, M., Simon, M., Urban, K., Lagally, M.G., Direct Determination of the Interaction between Vacancies on InP(110) Surfaces, Phys.Rev.Lett., 76, (№ 12), 18 March 1996.

[6] Gibbs, J., The Scientific Papers, 1, Dover, New York (1961).

[7] Heinrich, M., Ebert, Ph., Simon, M., Urban, K., Lagally, M.G., Temperature Dependent Vacancy Concentrations on InP(110) Surfaces, J. Vac. Sci. Technol. A. 13(3), May/Jun. 1995.

[8] Hörmander, L., The Analysis of Linear Partial Differential Operators, v. I-IV, Springer-Verlag, Berlin (1983).

[9] Israelishvili, J., Intermolecular and Surface Forces, Academic Press, London (1991).

[10] Jackson, J.D., Classical Electrodynamics, John Wiley & Sons, New York, (1962).

[11] Junqueira, Z., Carneiro, J., Kelly, R.O., Basic Histology, A. ZANCE Medical Book (1995).

[12] Landau, L., Lifschitz, S., Lectures on Modern Physics, vol. VIII Electrodynamics of Solids, Nauka (Moskow) 1982 (in Russian).

[13] Radoev, B., Boev, T., Avramov, M. Electrostatics of Heterogeneous Monolayers, Adv. In Colloid and Interface Sci., 114-115 (2005) 93-101.

Part 5

Nanoelectronics

5

Nanowires: Promising Candidates for Electrostatic Control in Future Nanoelectronic Devices

Dura Julien[1,2], Martinie Sébastien[2], Munteanu Daniela[2], Triozon François[1], Barraud Sylvain[1], Niquet Yann-Michel[3] and Autran Jean-Luc[2]
[1]CEA-LETI MINATEC and [3]CEA-UJF,
Institute for Nanosciences and Cryogenics (INAC), Grenoble,
[2]IM2NP-CNRS, UMR CNRS 6242, Marseille,
[1,2,3]France

1. Introduction

The microelectronics activity regroups the study, design, and manufacturing of very small electronic components. These devices are essentially based on interconnected transistors, sort of "switches" which allow controlling the electric current, and are made of semiconductor materials. Depending on the voltage applied to its "gate" electrode, a transistor is in ON state (high current) or OFF state (smallest possible current and low power consumption). Since the invention of the first transistor in 1948, technological progress allowed miniaturizing drastically electronic circuits, and the industry grew fast up to now. For example, the first microprocessor of INTEL (the "4004") contained 2300 transistors while the Pentium 4 in the early 2000's got 55 millions of transistors and the dual core more than 150 millions. To have a clear idea on the fast growing of this industry, in the 60's and 70's, the number of transistors in integrated circuits was doubled every year. Since the 80's, the standard rule is a factor 2 every 18 months. This evolution is more known as the "Moore's law". Of course, such an industry implies several companies. Microelectronics is become very competitive in performances as well as for economical aspects. The price of 1 million of transistor was 75000 € in 1973, while it was of 6 cents in 2000 then 0.5 cent in 2005. The common objective in microelectronics is so to go ahead with the improvement of transistor in all aspects (electronic performances and economical).

To follow this endless race, the well-known concept of downscaling is required consisting in continuously shrinking the geometrical dimensions of the transistor. However, for small device length, the electrostatics of the device is affected, which degrades the control of the electric current. So, to keep the performances under control, the device architectures have evolved by, for example, improving the gate (controlling electrode) or using thin-film transistor. This article focuses on MOSFETs (Metal-Oxide-Semiconductor Field-Effect Transistor) made on silicon, since it is the technology used since decades for microprocessors. The main part of the MOSFET is its semiconducting "channel" coupled to conducting "source" and "drain" regions, and surrounded by one or several gate electrodes which control the current through the channel. The gate is separated from the channel by a thin insulating oxide. Figure 1 shows transmission electron microscopy (TEM) images of several MOSFET architectures (a) and schematics of these devices and of their potential for channel length reduction (b).

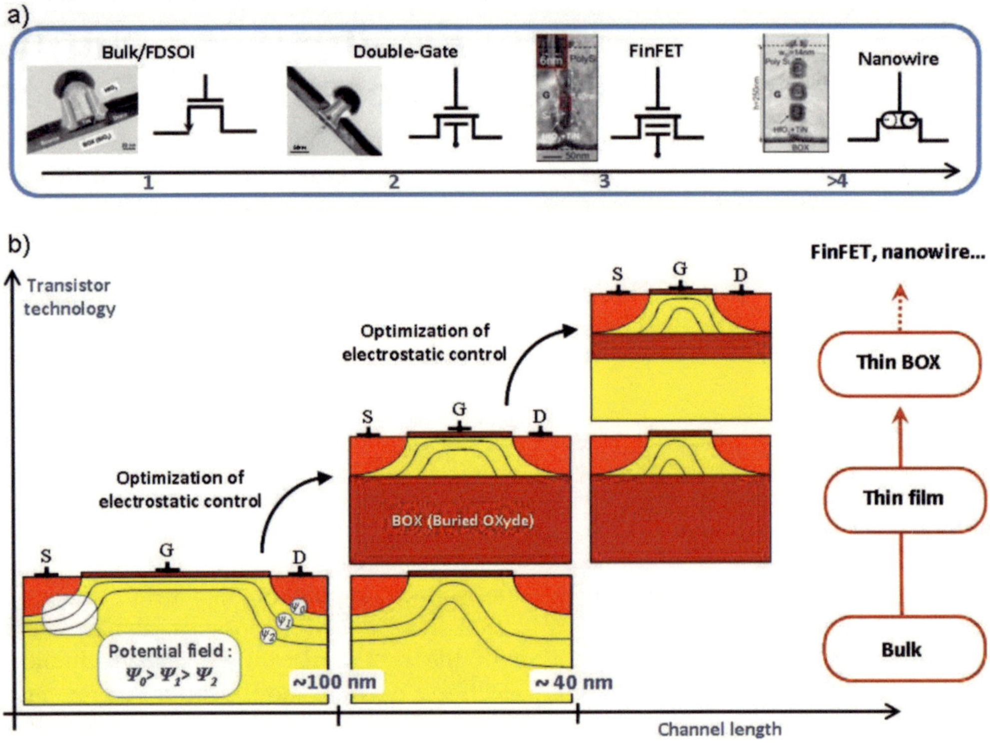

Fig. 1. a) Different MOSFET architectures observed by TEM (Fully-Depleted Silicon on Insulator FDSOI (Barral & al., 2007a), Double Gate (Barral & al., 2007b), finFET (Dupre & al., 2008) and stacked nanowires (Dupre & al., 2008)); b) schematics of the downscaling concept.

The essential parameter used to analyse the electrostatic behaviour (so to compare MOSFET architectures) is the natural length λ (Collinge, 2007). It represents the perturbation induced by the transistor source and drain junctions on the gate control. Numerical simulations establish that a device is relatively free of electrostatic perturbations if λ has a value smaller than 5-10 times the gate length.

$$\lambda = \sqrt{\frac{\varepsilon_{si}.t_{si}.t_{ox}}{N.\varepsilon_{ox}}} \tag{1}$$

where ε_{si} and ε_{ox} are the silicon and oxide permittivity, t_{si} and t_{ox} the silicon and oxide thickness and N represents the number of gates of the architecture. Thus, for a given value of silicon thickness and oxide (t_{si}=10 nm and t_{ox}=1.5 nm) the corresponding minimum length for the bulk, thin BOX FDSOI, and nanowire are 20 nm, 15 nm and 10 nm respectively. That is why ITRS recommends nanowires for technology node sub-22nm (International

Technology Roadmap of Semiconductor [ITRS], 2009) and regarding the advanced processing technologies, the literature provides a wide range of devices based on nanowires, stacked (Dupre & al., 2008), twin (Hwi Cho & al., 2007) or single Ω-FET nanowires (Tachi & al., 2009). In the following, a complete study of the electrostatics of nanowire MOSFETs is performed including all the ultimate physical phenomena which can occur in future electronic devices.

2. The electronic structure of silicon nanowires

Standard silicon layers used in microelectronics are crystallographic. Silicon atoms are disposed in a periodical lattice similar to the diamond structure: each atom is tetrahedrally bonded to its four neighbours (see figure 2). The cubic unit cell parameter a_0 equals 5.43 Å, corresponding to an interatomic distance of 2.34 Å. Ideal silicon nanowires are thus periodic along their axis, and the length L of their unit cell depends on the crystallographic orientation:

- $L=a_0$ for <100> oriented nanowires,
- $L=a_0/\sqrt{2}$ for <110> oriented nanowires,
- $L=a_0\sqrt{3}$ for <111> oriented nanowires.

The orientation and diameter of the nanowire determines its electronic structure, from which result its electrical and optical properties. In the following of this work, we will consider cylindrical nanowires oriented along the <100> axis, as the one represented in figure 2.

The electronic structure of bulk silicon is expressed by the dispersion relations $E_n(k)$, which give the energy of an electron wavefunction with wavevector k in band n. A schematic of low energy electrons in the conduction band is shown in figure 3. It represents the iso-energy surfaces in the first Brillouin zone (wavevector space). For conduction bands, we can count six energy minima, named the six "Δ valleys" of bulk silicon. Each valley is characterized by an effective mass m_l along its orientation axis and m_t along its transverse directions. The longitudinal mass is m_l equal to $0.919m_0$ and m_t is equal to $0.196m_0$ where m_0 is the free electron mass.

In the following, we consider transport along the x-axis. So, projecting bulk valleys on the nanowire axis (x for the transport, y and z for the perpendicular direction), we can define two different valleys of the nanowire characterized by a conduction and a confinement masses. The valleys 1 and 2 of the figure 3 correspond to the longitudinal valley while the valleys 3, 4, 5 and 6 refer to the transverse valley. The table 1 gives the corresponding masses of the two nanowire valleys. The confinement mass of the transverse valleys is approximated by a "cylindrical mass", which preserves cylindrical symmetry in the calculations.

	Conduction mass	Confinement mass
Longitudinal valley	m_l	m_t
Transverse valley	m_t	$2.m_l.m_t/(m_l+m_t)$

Table 1. Definition of the longitudinal and the transverse masses for the <100> oriented nanowire.

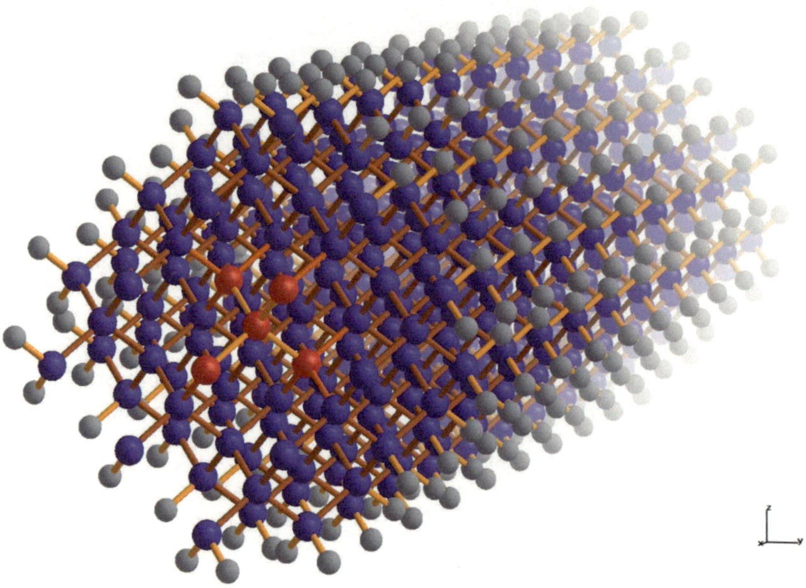

Fig. 2. Atomistic representation of a <100>-oriented nanowire. Blue: silicon atoms, grey: hydrogen atoms (necessary to passivate the surface in the tight-binding model), red: highlight of the tetrahedral structure of silicon. The nanowire is 1.5nm thick and 5nm long.

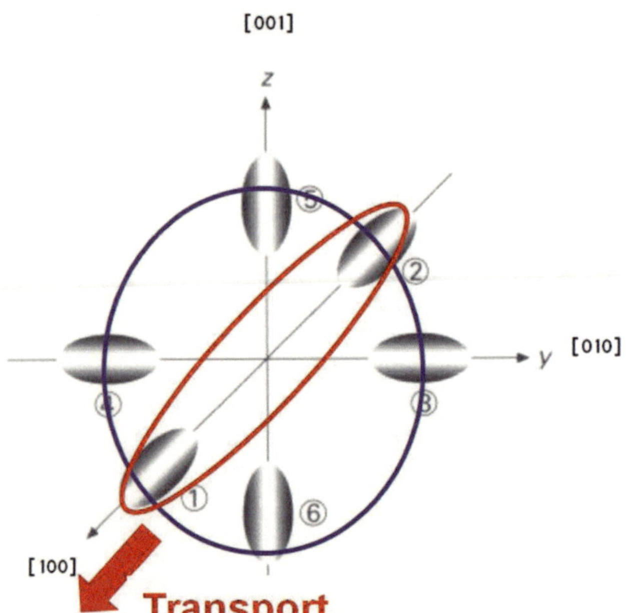

Fig. 3. Iso-energy surfaces of the conduction band of bulk silicon in wavevector space, and definition of longitudinal and transverse valleys for transport along a <100>-oriented nanowire.

This effective mass approach is only valid for thick nanowires (diameter > 5 nm) and low electron energy (a few tenths eV). The band structure of silicon is described more accurately by atomistic models, which allow modelling thinner nanowires. A tight-binding model is used here. It consists in developing the wavefunctions on an atomic orbital basis set. The sp^3 model developed by (Niquet & al., 2000) is used here and in previous studies (see section 3.4). This model is fitted on first principles calculations based on the density functional theory (DFT) and so-called "GW" corrections for the bandgap. It contains one *s* orbital and three *p* orbitals per silicon atom and describes accurately electron (conduction band) and hole (valence band) dispersion relations. When studying nanostructures, the surface is passivated with hydrogen atoms (see figure 2). This model choice is due to the lack of a convenient tight-binding model for the Si/SiO_2 interface. Passivation avoids unrealistic surface states and should not modify much the electronic structure of the nanowire. The tight-binding calculations are performed with the code TB_Sim (TBSIM, 2011), which solves the Schrödinger equation in nanostructures containing up to 10^7 atoms (Niquet & al., 2006). For thin nanowires (diameter < 5 nm), the obtained electronic structure differs from the effective mass calculation (see section 3.4). TB_Sim also allows Poisson-Schrödinger calculations, which give the repartition of the charge density in the nanowire under the influence of the gate voltage. Again, corrections of the effective mass approach are needed for thin nanowires.

3. Modelling of the electrostatics in MOSFETs based on nanowires

3.1 Definition of the threshold voltage

As said previously, the MOSFET transistor is defined by two states (ON or OFF) depending on the voltage applied at the gate. In fact, this polarization creates an electric field in the active region of the transistor which makes carriers concentrate near the interface with the oxide. Increasing the gate voltage, conduction bands start to fill with carriers from lower energy bands to higher energy bands up to saturation. In this regime, the semiconductor is then analog to a metal and forms a conduction layer between contacts (source and drain) of the transistor. Commonly, the threshold voltage is so defined as the frontier of the two states of the transistor and represents its capacity to switch from one state to the other. It is essentially dependent on the electrostatic characteristics. That is why, it is necessary to fully describe the potential ψ everywhere in the device active region. For this purpose, the Poisson equation, given here in cylindrical coordinates, is solved:

$$\frac{d^2\psi}{dr^2}+\frac{1}{r}.\frac{d\psi}{dr}=\frac{q}{\varepsilon_{Si}}(N_A+n) \tag{2}$$

where N_A is the channel doping, n is the electron density, ε_{Si} the silicon permittivity, and q the elementary charge. Note that n depends on ψ via the Fermi-Dirac occupation of electronic states.

Two approaches can be used to obtain the solution of such an equation. The first one is the double integration solving (Jimenez & al., 2004; Yu & al., 2007); however the solution is not totally analytical which is not convenient in our case. The second approach is to make an assumption on the potential description along the nanowire radius. In the following, we assume a parabolic potential along the radius of the nanowire cross-section described as:

$$\psi(r,x) = \beta_1 + \beta_2.r + \beta_3.r^2 \tag{3}$$

where the parabolic terms β_i are x-dependent functions.

To define these terms, the general expression of the potential (eq. 3) is injected in boundary conditions specific to nanowires (eq. 4): the potential at the position r=D/2 is equal to the potential at the position r=-D/2 (symmetry condition) and is defined as the surface potential ψ_s:

$$\psi(D/2,x) = \psi(-D/2,x) = \psi_s(x) \tag{4}$$

Equation (3) becomes:

$$\psi(r,x) = \psi_s(x) - \beta_3.\frac{D^2}{4} + \beta_3.r^2 \tag{5}$$

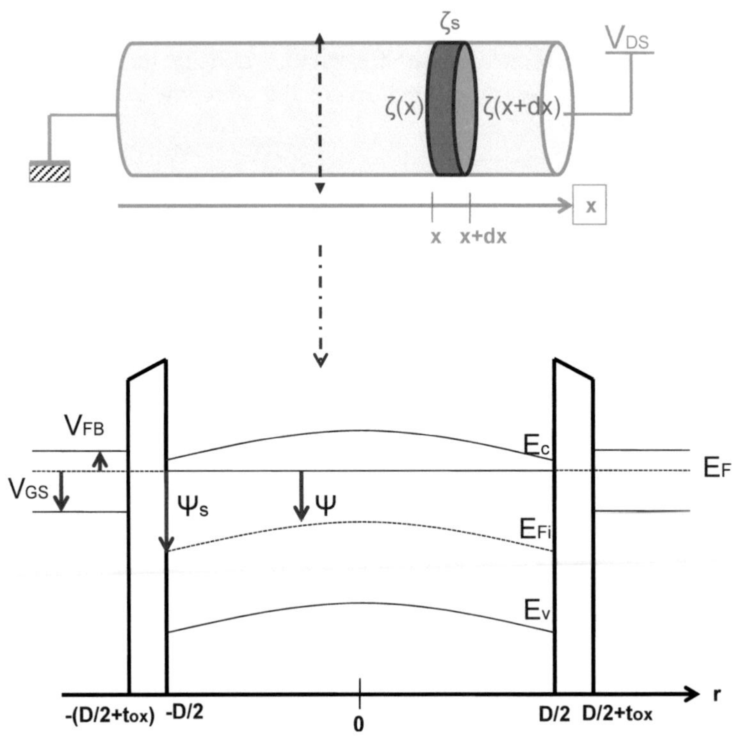

Fig. 4. (a) Schematics of a nanowire device (for a better view the gate oxide and material are not shown), indicating the specific area used in the Gauss law. (b) Band diagram along a vertical cut-line in the nanowire.

The term β_3 is found by including (5) into the Poisson equation (2) and integrating along the nanowire radius from 0 to D/2:

$$\beta = \beta_3 = \frac{q.N_A}{4\varepsilon_{Si}} + \frac{Q_{i,lin}}{\varepsilon_{Si}.D} \tag{6}$$

where $Q_{i,lin}$ is the charge integrated along the nanowire radius and ψ_s is the surface potential.

It is important to note that the parabolic assumption is valid at threshold, but could lose its validity in other operation regimes (for example in the strong inversion regime). Figure 4(a) shows the schematics of the nanowire with the longitudinal polarization (V_{DS}) along the x-axis (transport). Figure 4(b) shows the band diagram along a transverse cutline in the nanowire for an applied gate voltage V_{GS}. The potentials are defined with respect to the intrinsic Fermi level as illustrated in figure 4(b).

The starting point of the threshold voltage modelling is the boundary condition at the Si/SiO_2 interface:

$$V_{GS} - V_{FB} = \frac{\varepsilon_{Si}}{C_{ox}}.\zeta_S + \psi_S + \phi_F \tag{7}$$

where V_{FB} is the flat-band voltage, ζ_S is the electric field at the interface, ϕ_F is the Fermi potential and C_{ox} is the oxide capacitance in cylindrical coordinates expressed in (Dura & al., 2010).

In our case, we consider the threshold voltage defined as the gate voltage for which the inversion charge reaches its threshold value fixed to (Munteanu & al., 2005):

$$Q_{ith,lin} = \frac{kT}{q}.C_{ox} \tag{8}$$

where k is the Boltzmann constant and T is the temperature. Under this condition, the surface potential reaches its threshold value, called $\psi_{s,th}$. The threshold voltage is then obtained as:

$$V_{th} = V_{FB} + \frac{\varepsilon_{Si}}{C_{ox}}.\beta.D + \psi_{s,th} + \phi_F \tag{9}$$

with

$$\beta = \frac{q.N_A}{4\varepsilon_{Si}} + \frac{kT}{q}\frac{C_{ox}}{\varepsilon_{Si}.D} \tag{10}$$

In equation (9), only the surface potential is unknown and has to be modeled taking into account the physical phenomena specific to nanowire MOSFETs described below: quantum confinement, short channel effect, and band structure effect.

3.2 Quantum mechanical confinement (QE)

In silicon nanostructures such as nanowires, the wavefunctions related to the different valleys are modified and kinetic energy is quantized along the confinement directions, leading to a set of energy subbands for each valley. Previous works highlight the necessity to consider quantum confinement in the transport modelling of planar architectures (Munteanu & al., 2005), for which the confinement is one-dimensional. For nanowire devices, quantum confinement is two-dimensional (leading to a 1D electronic gas) and its impact is expected to be stronger (Autran & al., 2005). Figure 5 shows the wavefunction in the first five subbands for the longitudinal valley and for two different nanowire diameters (5 and 10nm). These wavefunctions are calculated using an effective mass Schrödinger-Poisson solver (TBSIM, 2011). The associated energy represents the difference between each energy subband and the first subband level.

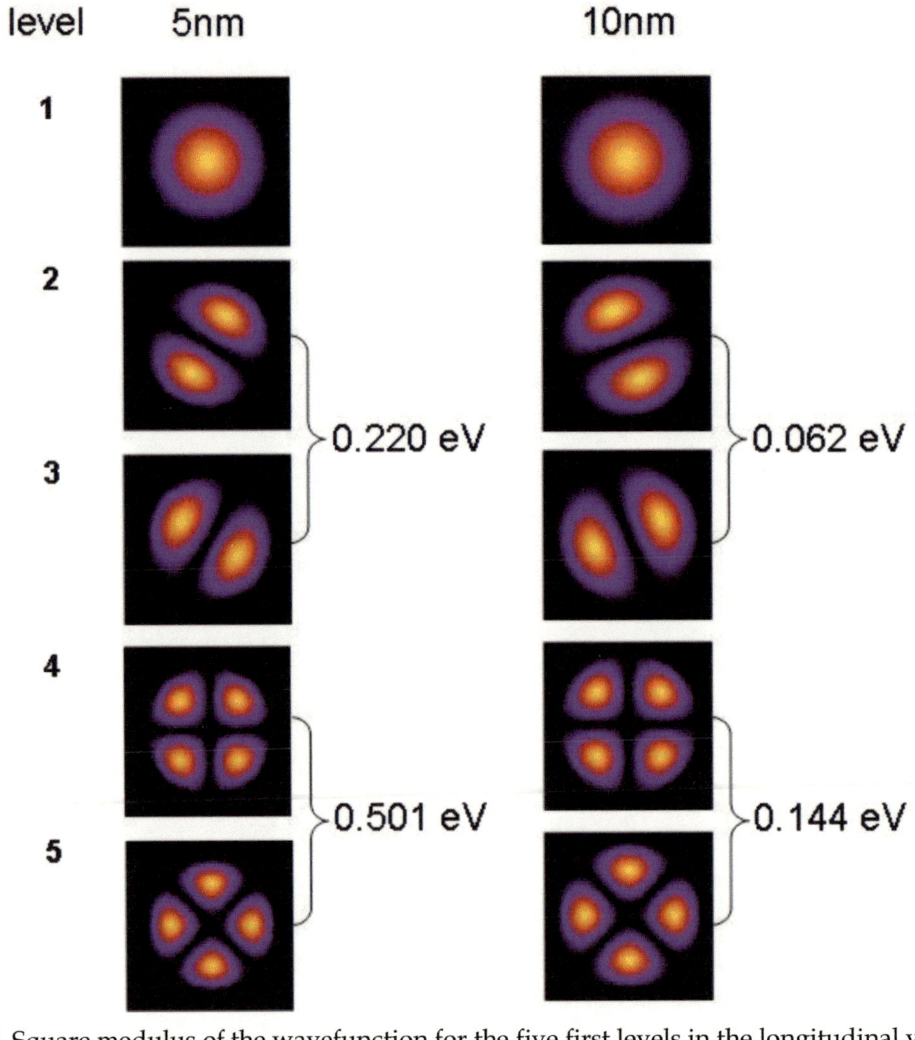

Fig. 5. Square modulus of the wavefunction for the five first levels in the longitudinal valley of a 5 and 10nm nanowire diameter. Energy increase of each subband with respect to the first level.

To extract numerically the impact of the diameter on the quantization of carriers, the same calculation has been done for the tranverse valley and for different nanowire diameters. We can note that for low diameters, the quantized levels are higher. So, regarding the targeted MOSFET downscaling to a few nanometers, quantum-mechanical effects have to be included in the previous analytical model of threshold voltage in order to assess their impact up to circuit performances.

For this purpose, the 1D quantum charge integrated along the radius $Q_{i,lin}$ has to be used:

$$Q_{i,lin} = \sum_j \sum_i n(E_c, i, j) \tag{11}$$

with

$$n(E_c, i, j) = 2 \int_{E_c}^{+\infty} \rho_{1D}(E, E_c).f(E).dE \tag{12}$$

where the factor 2 accounts for the number of equivalent valleys, j is the valley index, i is the subband index, $Ec = E^i_j$ is the subband bottom energy, $\rho_{1D}(E,E_c)$ is the 1D density-of-states of the subband, and $f(E)$ is the Fermi-Dirac distribution function:

$$\rho_{1D}(E, E_c) = \frac{1}{\pi}.\left(\frac{2.mj}{\hbar^2}\right)^{1/2}.\frac{1}{\sqrt{E - E_c}} \tag{13}$$

$$f(E) = \frac{1}{1 + \exp\left(\dfrac{E - E_F}{kT}\right)} \tag{14}$$

where m_j is the 1D density-of-states effective mass in valley j and E_F is the Fermi level. The general expression of the charge is then:

$$Q_{i,lin} = q \sum_j \sum_i \left(\frac{1}{\pi}.g(j).\sqrt{\frac{2m_j}{\hbar^2}}\right).\sqrt{\frac{kT}{q}}.\int_0^{\infty} \frac{r^{-0.5}}{1 + e^{r - \frac{q}{kT}\left(E^i_j - \frac{Eg}{2} - \psi_S\right)}}.dr \tag{15}$$

Under non-degenerate condition (valid at threshold), the Fermi-Dirac distribution can be approximated by a simple exponential (corresponding to the Boltzmann distribution). The expression of the charge becomes (Autran & al., 2004):

$$Q_{i,lin} \approx Q^*.e^{\frac{q.\psi_S}{kT}} \tag{16}$$

with

$$Q^* = q \sum_j \sum_i \left(\frac{1}{\pi}.g(j).\sqrt{\frac{2m_j}{\hbar^2}}\right).\sqrt{\frac{kT}{q}}.\sqrt{\pi}.e^{-\frac{q}{kT}\left(E^i_j + \frac{Eg}{2}\right)} \tag{17}$$

where Eg is the silicon bandgap, and g(j) is the degeneracy of the valley j (equal to 2 for each valley). The charge is then obtained by a sum over the different silicon valleys (index j) and a sum over all quantized levels (index i) of each valley (in practice limited to 5). The quantum energy levels are needed in (17) and have to be calculated analytically. For a cylindrical cross-section of the nanowire, the analytical expressions of the transversal (index t) and longitudinal (index l) quantum energy levels are given by (Baccarani, 2008):

$$\begin{aligned} E_t^i = E_{1,2}^i &= \frac{(\alpha.\hbar.i)^2}{4.q.(D/2)^2}.(\frac{1}{m_t}+\frac{1}{m_l}) \\ E_l^i = E_3^i &= \frac{(\alpha.\hbar.i)^2}{2.q.(D/2)^2}.\frac{1}{m_t} \end{aligned} \tag{18}$$

where α is a numerical parameter (Baccarani, 2008). We can note a good agreement between (18) with α=2.1 and a self-consistent cylindrical 1D Schrödinger-Poisson solver (ATLAS, 2010) for the first energy level (transversal and longitudinal) (Dura & al., 2010).

From equation (16), the surface potential at threshold voltage is given by:

$$\psi_{s,th} = \frac{kT}{q}.\ln(\frac{Q_{ith,lin}}{Q^*}) \tag{19}$$

with $Q_{ith,lin}$ the inversion charge at threshold defined by (8).

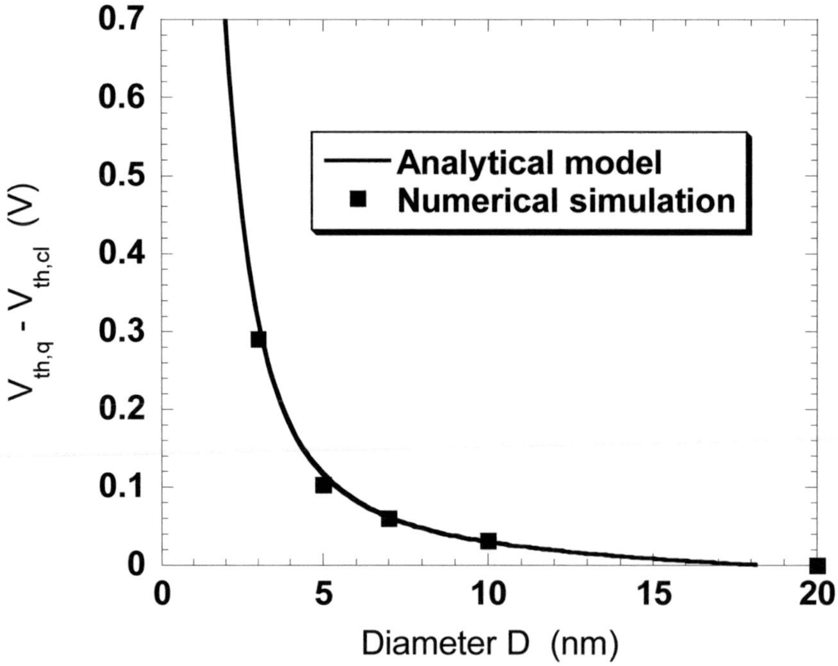

Fig. 6. Threshold voltage shift between quantum (Vth,q) and classical (Vth,cl) approaches versus nanowire diameter in long channel transistors. Comparison between the analytical model and data obtained from Schrödinger-Poisson numerical solving (ATLAS, 2010).

Including this expression in (9), the quantum threshold voltage for long channel transistor is easily obtained. Figure 6 plots the difference between quantum and classical threshold voltage versus the nanowire diameter; as expected, this difference increases when reducing the nanowire diameter, due to a stronger quantization of carrier energy when the nanowire diameter is reduced. Figure 6 also shows a very good agreement between the analytical model and data obtained by a Schrödinger-Poisson numerical solving (ATLAS, 2010).

3.3 Short-Channel Effect (SCE)

As said in the introduction, the downscaling of transistor is required. That is why the gate length is continuously reduced. However, from a certain dimension, the transistor junctions (source and drain) have an impact on the electrostatic control of the device. Previously 1D (only the gate voltage), it becomes 2D because gate and drain polarizations compete to control the device. And this strongly affects the device characteristics. Thus, MOSFET architectures are considered to be impacted by the short channel effect when the channel length is the same order of magnitude as the depletion-layer widths of the source and drain junction. The main result is the modification of the threshold voltage (or the loss of electrostatic control) due to the shortening of the channel length. It is attributed to two phenomena: SCE (Short Channel Effect) and DIBL (Drain Induced Barrier Lowering). The first one is coming from the superposition of the depletion-layer widths of the source and drain junction. The second phenomenon is a secondary effect on the charge sharing due to higher drain voltage. Nanowire transistors being expected for ultimate technology node, consideration of short channel effect is required in a realistic modeling of this architecture.

To fully describe 2D electrostatic effects (SCE and DIBL) in short channel devices, we propose a full analytical model describing the threshold voltage impacted by SCE and DIBL. The x-dependence (transport direction) of the surface potential has to be know. Applying the Gauss law on a nanowire slice as illustrated in figure 4(a), the following equation is obtained (Munteanu & al., 2005):

$$-\zeta(x).\pi.\frac{D^2}{4}+\zeta(x+dx).\pi.\frac{D^2}{4}-\zeta_S(x).\pi.D.dx=-\frac{q.N_A.\pi.D^2}{4.\varepsilon_{Si}} \tag{20}$$

where ζ is the electric field expressed by:

$$\zeta(x)=-\eta.\frac{d\psi_S(x)}{dx} \tag{21}$$

η is a fitting parameter which models the lateral electric field variation (Banna & al., 1995). It depends on the channel doping, the channel length, the nanowire diameter and the polarization. An empirical formula for η, obtained from numerical simulations, will be presented below to include these both effect. Introducing (21) in (20), we find a second order equation for the surface potential:

$$\frac{d^2\psi_s}{dx^2}-4\frac{C_{ox}}{\eta.\varepsilon_{Si}.D}.\psi_s=\frac{2}{\eta.\varepsilon_{Si}.D}[q.N_A.\frac{D}{2}-2C_{ox}(V_{GS}-V_{FB}-\phi_F)] \tag{22}$$

The solution of this equation is given by (Munteanu & al., 2005):

$$\psi_s=K_1.e^{\gamma.x}+K_2.e^{-\gamma.x}-\frac{K_3}{\gamma^2}$$
$$\gamma=\sqrt{\frac{4.C_{ox}}{\eta.\varepsilon_{Si}.D}} \tag{23}$$

where K_1, K_2 and K_3 are functions resulting from the Poisson equation solving:

$$K_1 = \frac{(1-e^{-\gamma.L_c}).(V_b + K_3/\gamma^2) + V_{DS}}{2.sh(\gamma.L_c)} \tag{24}$$

$$K_2 = -\frac{(1-e^{+\gamma.L_c}).(V_b + K_3/\gamma^2) + V_{DS}}{2.sh(\gamma.L_c)} \tag{25}$$

$$K_3 = \frac{2}{\eta.\varepsilon_{Si}.D}[q.N_A.\frac{D}{2} - 2C_{ox}(V_{GS} - V_{FB} - \phi_F)] \tag{26}$$

where V_b is the built-in potential depending on the channel doping N_A, the source/drain doping N_{SD} and the intrinsic carrier density n_i as:

$$V_b = \frac{kT}{q}.\ln\left(\frac{N_A.N_{SD}}{n_i^2}\right) \tag{27}$$

The position where the surface potential is minimum x_{min} and the value of ψ_S ($\psi_{s,min}$) at the position x_{min} are obtained by forcing the first derivative of equation (23) to be equal to zero (Munteanu & al., 2005):

$$x_{min} = \frac{1}{2.\gamma}.\ln(\left|\frac{K_2}{K_1}\right|) \tag{28}$$

$$\psi_s = \psi_{s,min} = 2\sqrt{K_1.K_2} - K_3/\gamma^2 \tag{29}$$

We assume that the transistor switches on when $\psi_{s,min} = \psi_{s,th}$. By inserting this expression in the general expression of the threshold voltage (9), we obtain:

$$V_{th} = V_{FB} + \frac{\varepsilon_{Si}}{C_{ox}}.\beta.D + \phi_F - K_3/\gamma^2 + 2\sqrt{K_1.K_2} \tag{30}$$

We can, by analogy to (Suzuki & al., 1996), distinguish two different terms. The first one (independent from the channel length) refers to the long-channel threshold voltage $V_{th,long}$; the second term, which tends to zero for long channel length, represents the threshold voltage roll-off and describes the impact of SCE/DIBL:

$$V_{th,long} = V_{FB} + \frac{\varepsilon_{Si}}{C_{ox}}.\beta.D + \phi_F - K_3/\gamma^2 \tag{31}$$

$$\Delta V_{th} = -2\sqrt{K_1.K_2} \tag{32}$$

where the variation of the threshold voltage is defined as:

$$\Delta V_{th} = V_{th} - V_{th,long} \quad (33)$$

At threshold, $V_{GS}=V_{th}$ in (26) and, considering (33), K_3 will depend on ΔV_{th}. Moreover K_1 and K_2 depend on K_3, they will also depend on ΔV_{th}. Then, developing (33) leads to a second order equation of ΔV_{th} as:

$$A.\Delta V_{th}^2 + B.\Delta V_{th} + C = 0 \quad (34)$$

with

$$\begin{cases} A = sh(\gamma.L_C)^2 + 2.[1 - ch(\gamma.L_C)] \\ B = 2.[1 - ch(\gamma.L_C)].[2.H_4 + V_{DS}] \\ C = 2.H_4.[1 - ch(\gamma.L_C)].[H_4 + V_{DS}] + V_{DS}^2 \\ D = V_b - V_{th,long} + V_{FB} + \phi_F \end{cases} \quad (35)$$

Finally, the threshold voltage roll-off is the solution of (34) given by:

$$\Delta V_{th} = \frac{-B + \sqrt{B^2 - 4.A.C}}{2.A} \quad (36)$$

We can note that to find this term, we have to take into account the long-channel threshold voltage (see coefficient D in eq.35) which includes the dependence on the quantum confinement. Consequently, this model includes both the impact of quantum confinement on the long channel and the threshold voltage roll-off. The most common model (such as references (Banna & al., 1995; Suzuki & al., 1996)) does not include the effect of quantum confinement on the evolution of the short channel effect in analytical modeling which becomes dominant in nanoscale device such as nanowire (this aspect will be detailed later in paragraph 4).

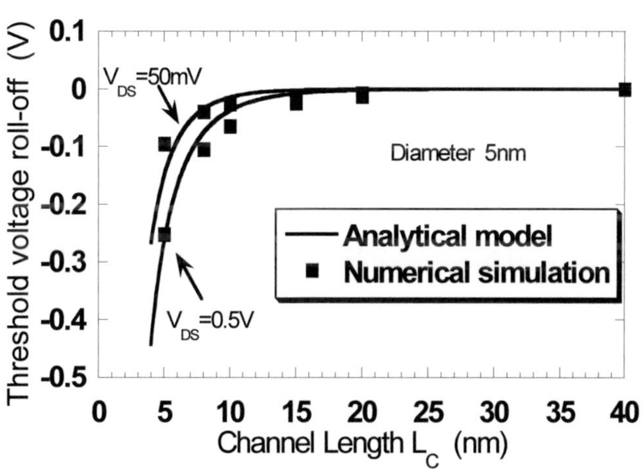

Fig. 7. Threshold voltage roll-off versus channel length for low (V_{DS}=50mV) and high (0.7V) drain voltage obtained by the analytical model and TCAD simulations for a 5 nm nanowire diameter; t_{ox}=1 nm.

The threshold voltage roll-off is represented in figure 7 for nanowire diameters of 5 nm (calculated in the classical case, i.e. without quantum confinement). TCAD numerical simulations have been done for a cylindrical structure using a drift-diffusion model in order to obtain drain current characteristics as a function of the gate voltage. The threshold voltage is extracted from these current-voltage characteristics using the classical constant current method. The threshold voltage data extracted from TCAD simulations for different diameters have been used to derive an empirical expression of the parameter η including the dependence on the channel length, nanowire diameter and drain to source voltage:

$$\eta = \frac{D}{f_0 + D} + V_{DS}.\left[f_1.L_C + f_2.L_C^2\right] \quad (37)$$

where f_0, f_1 and f_2 are constant fitting parameters, calibrated on numerical simulations. Equation (37) is valid for a wide range of nanowire diameter (down to 2 nm) and channel lengths (down to channel length equal to the nanowire diameter). The results in figure 7 show a good agreement between the analytical model and threshold voltage data obtained from TCAD simulations.

3.4 Band Structure Effect (BSE)

Advanced atomistic numerical simulations (Neophytou & al., 2008; Niquet & al., 2000, 2006; Sarrazin & al., 2009; Nehari & al., 2006) have shown that a strong reduction of the silicon thickness impacts the material properties by modifying the band structure. Indeed, the dimensions targeted in ultra-scaled devices are those of a few tens atomic layers (several nanometers). At these dimensions, the electronic properties differ from the calculations shown in section 3.2 and based on the bulk effective masses. In (Sarrazin & al., 2009), atomistic tight-binding (TB) Schrödinger-Poisson simulations have been performed for the case of [001] oriented silicon nanowire in order to highlight the variation of the band structure with the nanowire diameter. The code TB_Sim (TBSIM, 2011) has been used with a sp^3 tight-binding model (Niquet & al., 2000). Figure 8 shows the valence and the conduction bands for Si nanowire width of 2nm and 10nm. We can note that when thinning the silicon film the minimum of the conduction band is increased and the general shape of bands becomes smoother (Sarrazin & al., 2009). However, the bandgap increase is smaller than the effective mass result of section 2.4.

In order to include these modifications in the previous threshold voltage modelling, analytical expressions of parameters affected by the band structure effect (band gap and effective masses) are proposed here. Diameter-dependent analytical functions (fitted on numerical simulations as illustrated in figure 9) are found for the bandgap and effective masses (inspired from (Niquet & al., 2000):

$$E_g = E_{g,bulk} + \frac{K_1}{D^2 + A_1.D + B_1} \quad (37)$$

$$m_{t(l)} = m_{t(l),bulk} + \frac{K_2}{D^2 + A_2.D + B_2} \quad (38)$$

where A, B and K are fitting constants.

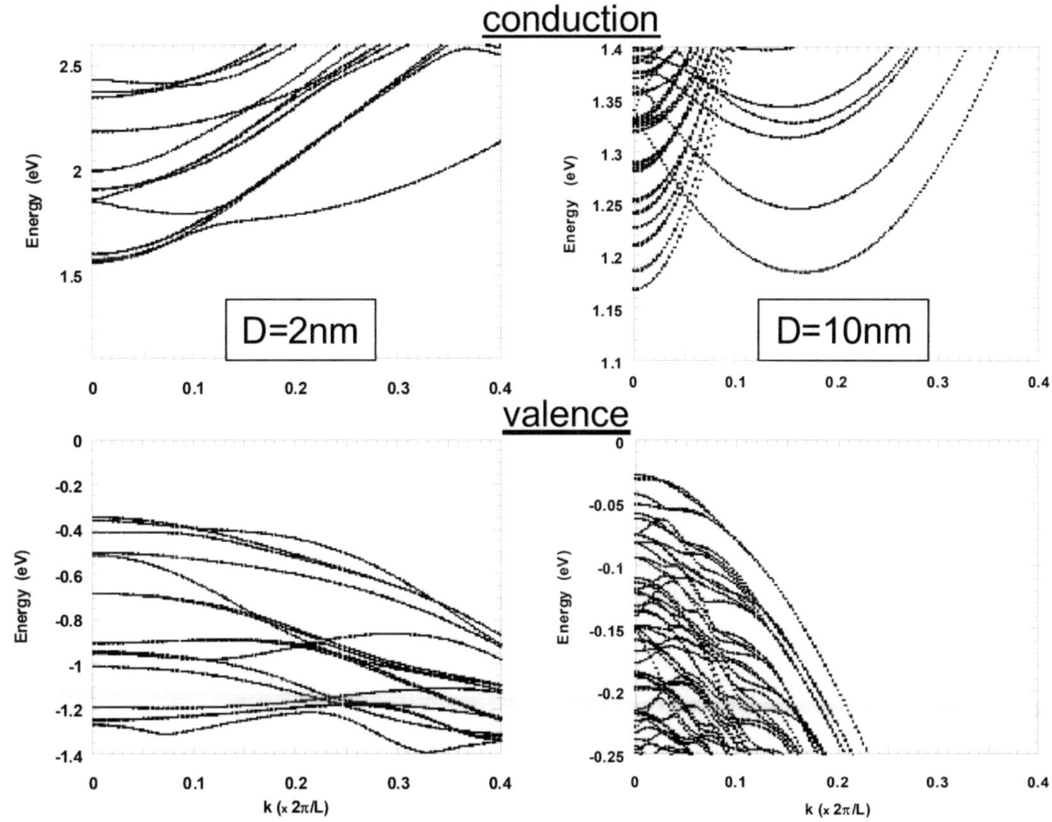

Fig. 8. Silicon nanowire band structure obtained with a tight-binding Schrödinger-Poisson solver (TBSIM, 2011) for two different diameters (2 and 10 nm). Up: Conduction band; down: Valence band.

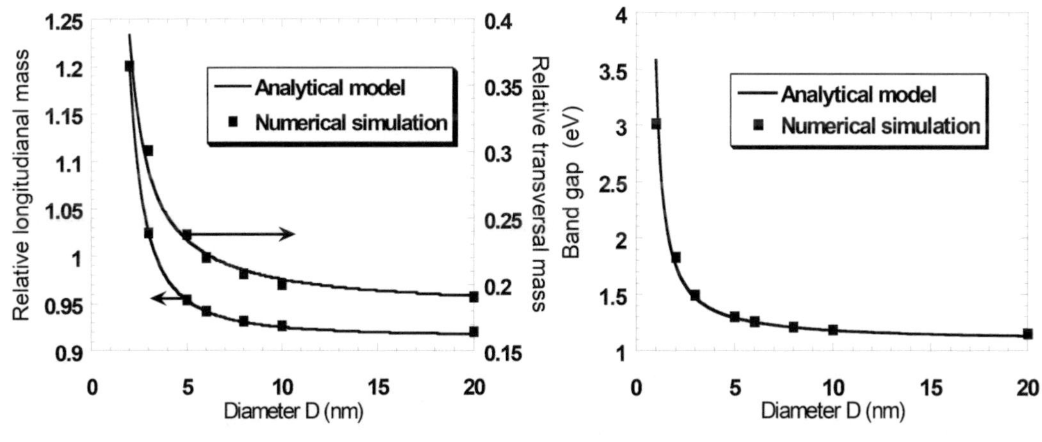

Fig. 9. Variation of the silicon band gap (a) and relative longitudinal and transversal masses (b) with respect to the silicon nanowire diameter. Comparison with atomistic simulations obtained in (Sarrazin & al., 2009).

4. Results and discussion

We have just presented the modeling of all the physical phenomena which affect the electrostatics of nanowire MOSFETs. In this part, the impact of each mechanism is assessed at different levels of interest: threshold voltage, drain current and small-circuits performance.

4.1 Impact on the threshold voltage

4.1.1 Long-channel transistors

Figure 10 shows the long-channel threshold voltage increase due to quantum confinement, without BSE (i.e., considering bulk value for the band gap and the conduction masses) and with BSE (i.e., considering equations (37) and (38)). The analytical model has been compared to both numerical simulations (Sarrazin & al., 2009) and experimental data (Suk & al., 2007).

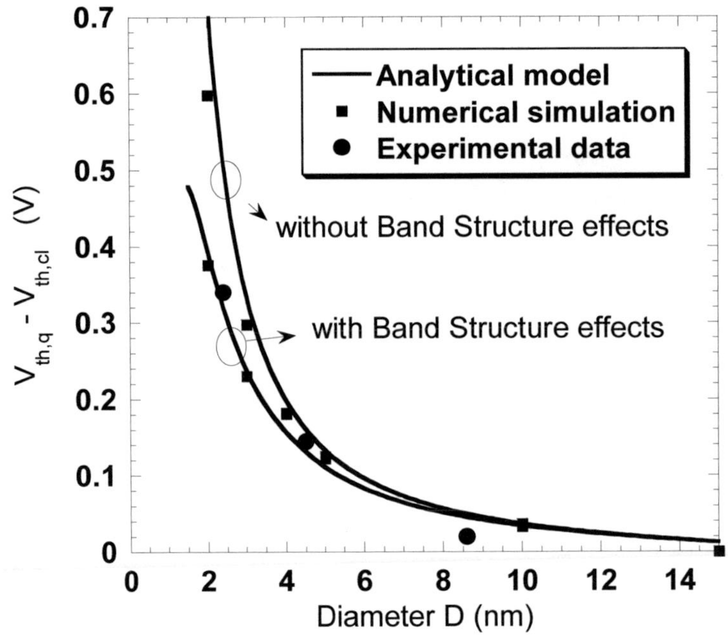

Fig. 10. Difference between quantum ($V_{th,q}$) and classical threshold voltage ($V_{th,cl}$) with and without BSE versus nanowire diameter (long channel transistors). Comparison between the analytical model, atomistic simulations (Sarrazin & al., 2009) and experimental data (Suk & al. 2007).

We can note that the band structure effects tend to limit the impact of the quantum confinement on the threshold voltage of the nanowire. This is coherent with equation (18). Increasing the effective masses, the quantum energy levels are lowered and the energy quantization decreases; then the quantum threshold voltage is lower. Figure 10 highlights the importance of considering BSE especially for thin film where the difference between threshold voltage with BSE and threshold voltage without BSE increases when reducing the nanowire diameter.

4.1.2 Short-channel transistors

In the following, we investigate the impact of the quantum confinement on SCE, then the impact of BSE on SCE. As stated previously, the threshold voltage roll-off depends on quantum confinement through the long-channel threshold voltage which includes quantum confinement effects.

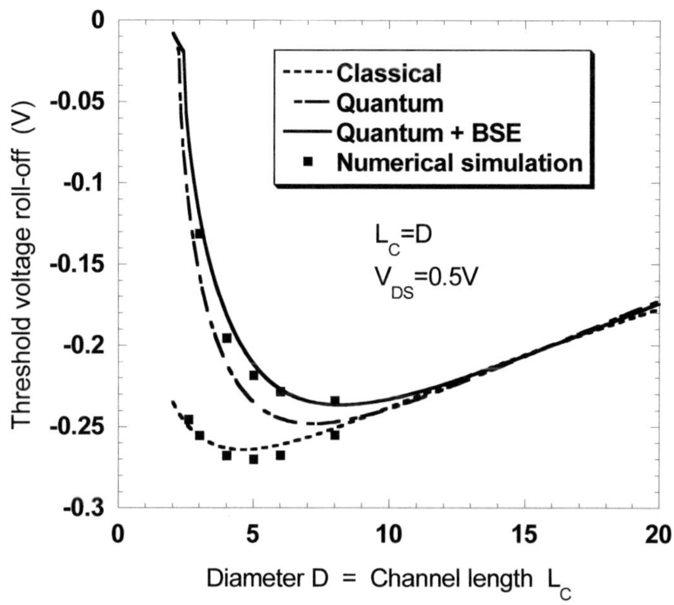

Fig. 11. Impact of band structure variations on SCE. Threshold voltage roll-off versus nanowire diameter for a channel length equal to the nanowire diameter. Comparison with data extracted from numerical simulations using a cylindrical Schrödinger-Poisson solver (Munteanu & Autran, 2003).

Figure 11 shows the impact of quantum effect and BSE on SCE as a function of the nanowire diameter. The curves plot the threshold voltage roll-off for a channel length equal to the nanowire diameter for the three approaches: classical (i.e., without quantum confinement and BSE), quantum without BSE, and quantum with BSE. Quantum threshold voltage obtained using the analytical model is validated in Fig. 11 with numerical simulation data obtained with a cylindrical Schrödinger-Poisson solver (Munteanu & Autran, 2003; Zervos & Feiner, 2004). We can note that the quantum confinement tends to limit SCE. This is due to the enhanced electrostatics control of the active area due to carrier energy quantization. As expected, the difference between quantum and classical approaches increases when reducing the nanowire diameter (due to the increase of energy quantum level for thinner films). When considering quantum confinement, the carrier energy is higher than for classical approach. That is why it is less affected by the longitudinal source to drain electric field, which generally strongly impacts the transistors performances at these channel length values. Moreover, figure 11 shows that BSE tend to amplify the impact of quantum effects on SCE: the threshold voltage roll-off is reduced when considering quantum confinement with BSE compared to the case when only quantum confinement is considered. For long channels, the threshold voltage decrease when considering BSE was due to the increase of the effective masses which lowered the quantized levels. In the case of short channels, the reduction of SCE when BSE are taken into account is the consequence of the band gap increase. A wider band gap means a higher energetic barrier, leading to a better electrostatics control which is less impacted by source-channel and drain-channel junctions when reducing the nanowire channel length. Moreover, we can note that below a certain diameter (depending on the modeling approach), the diameter thinning has a stronger impact on the threshold voltage roll-off than the channel length reduction. For the same channel length to diameter ratio, the threshold voltage roll-off is higher for D=5 nm than for D=2 nm. Indeed, for ultra-thin films, the quantization of carrier energy is very strong and the carrier concentration is mainly controlled by quantum confinement. In the case of D=2 nm, the strong electrostatic control due to the ultra-thin diameter completely overcomes the increase of SCE expected for these ultra-short channel lengths.

4.2 Impact on the injection velocity

Another parameter affected by the BSE is the thermal velocity which depends on masses along the transport direction. Indeed, in our case, for a transport along the (001) direction, the expression of thermal velocity is:

$$v_{th} = \sqrt{\frac{2.kT}{\pi.m_t}} \tag{39}$$

Figure 12 (Dura & al., 2011) shows the thermal velocity evolution with respect to the nanowire diameter. We can note a non-negligible reduction for ultra-thin nanowires up to a 20% decrease for D = 2 nm.

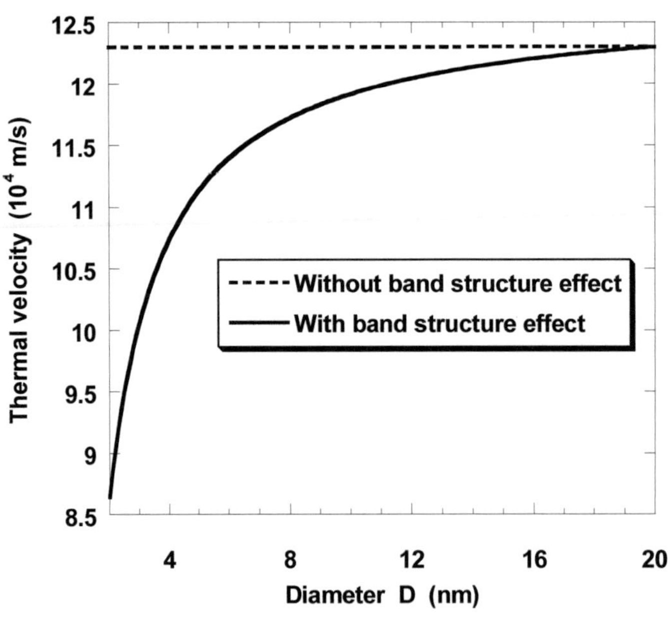

Fig. 12. Impact of the band structure effect on the thermal velocity with respect to the silicon nanowire diameter.

4.3 Impact on ballistic drain current of nanowire MOSFET

In previous works, we have demonstrated the analytical model of drain current in GAA nanowire MOSFETs in the ballistic transport regime (without interactions). We remind that this ballistic drain current is derived from the flux method initiated by McKelvey *et al* (McKelevey & al., 1961), doing a balance in the active region between the different carrier fluxes. In the degenerate case, the ballistic drain current is given by the following expression:

$$I_D = \pi.D.C_{ox}.(V_{GS} - V_t).\frac{\Im_0(\eta_F)}{\Im_{-1/2}(\eta_F)}.v_{th}.\left(\frac{1 - \dfrac{\Im_0(\eta_F - \frac{q.V_{DS}}{kT})}{\Im_0(\eta_F)}}{1 + \dfrac{\Im_{-1/2}(\eta_F - \frac{q.V_{DS}}{kT})}{\Im_{-1/2}(\eta_F)}}\right) \qquad (40)$$

where D is the nanowire diameter, v_{th} is the thermal velocity discussed above, V_{GS} is the drain to source voltage, V_{DS} is the drain to source voltage, C_{ox} is the oxide capacitance, η_F is the Fermi level, $\Im_0$ and $\Im_{-1/2}$ are the Fermi integral of order 0 and -1/2 respectively and V_t is the threshold voltage modeled above.

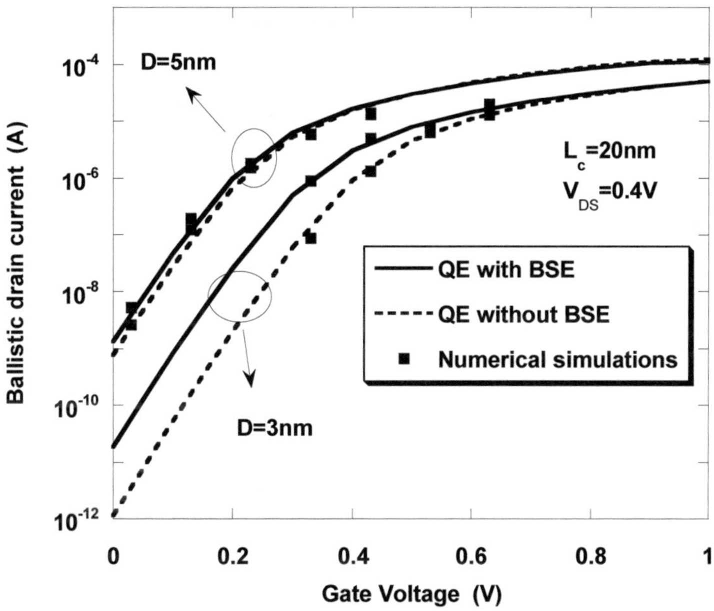

Fig. 13. Impact of the band structure effect on the ballistic drain current. Comparison with numerical simulations (deterministic Wigner equation solving (Barraud & al., 2009)).

Figure 13 shows the result at the device level for a long channel transistor and for two different nanowire diameters (3 and 5nm) (Dura & al., 2011). The ballistic drain current model is compared to numerical simulations based on a deterministic Wigner equation solver (Barraud & al., 2009). We can note a strong impact of BSE on the current in the subthreshold regime for 3nm-diameter due to the V_t variation while the ON-state current stays almost unchanged. From this graph, we can highlight the necessity to take into account the correction due to bands variations in the modeling if we expect to provide predictive devices performances. Indeed, at 3nm, the off-state current is increased by more than one decade when BSE is considered.

4.4 Impact on performances of small circuits based on nanowire MOSFETs

After implementation in a Verilog-A environment, the model presented above has been used to simulate a CMOS inverter and then a complete 11 stages-ring oscillator. In order to build-up the CMOS inverter a p-type nanowire MOSFET is considered symmetrically to the n-type transistor in the inverter setup. The impact of BSE can be addressed at the circuit level through the study of the commutation characteristics of the inverter or the oscillation frequency of a ring oscillator.

Figure 14 shows the input/output characteristics of the inverter for the classical case (i.e., without QE), with quantum effects (QE) and with band structure effects (QE+BSE). We can note that the inverter characteristic is more abrupt when considering only QE. Similarly to the results obtained for the threshold voltage, BSE tends to limit the impact of quantum confinement by smoothing the CMOS inverter switch.

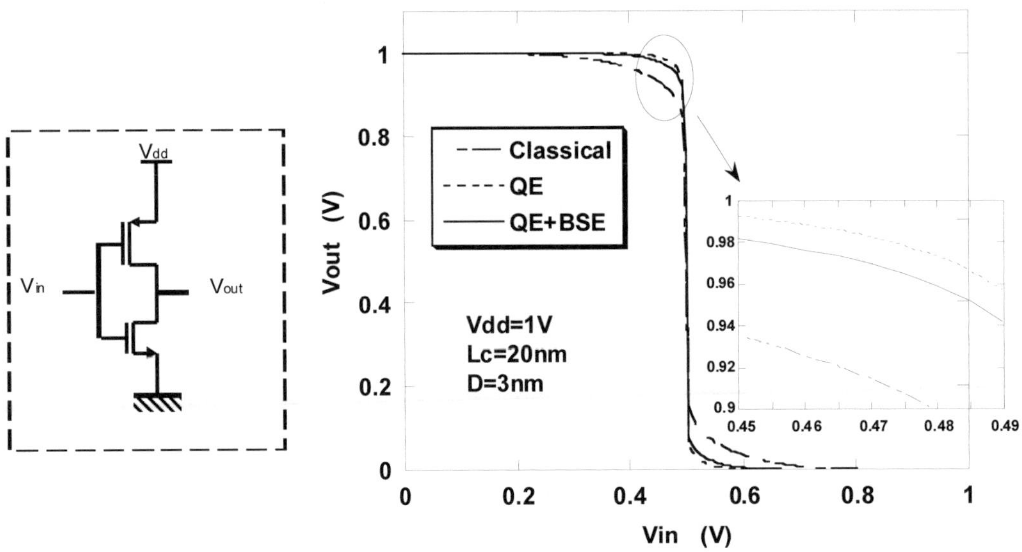

Fig. 14. Impact of BSE on the inverter characteristic. Comparison between classical (i.e, without QE), quantum (QE) and low dimensions effects (QE+BSE).

Regarding the ring oscillator, the results seem opposite to the inverter case. The better performances are for the classical case and introducing quantum confinement reduces the oscillation frequency. This is due to the fact that the ring-oscillator frequency is directly proportional to the ON-state current in strong inversion regime Vdd=1.5V (far from the threshold voltage). QE increases V_{th} and consequently reduces the current. The injection velocity also impacts directly the current and then the oscillation frequency is affected. The result is a reduction of the oscillation frequency when BSE are taken into account.

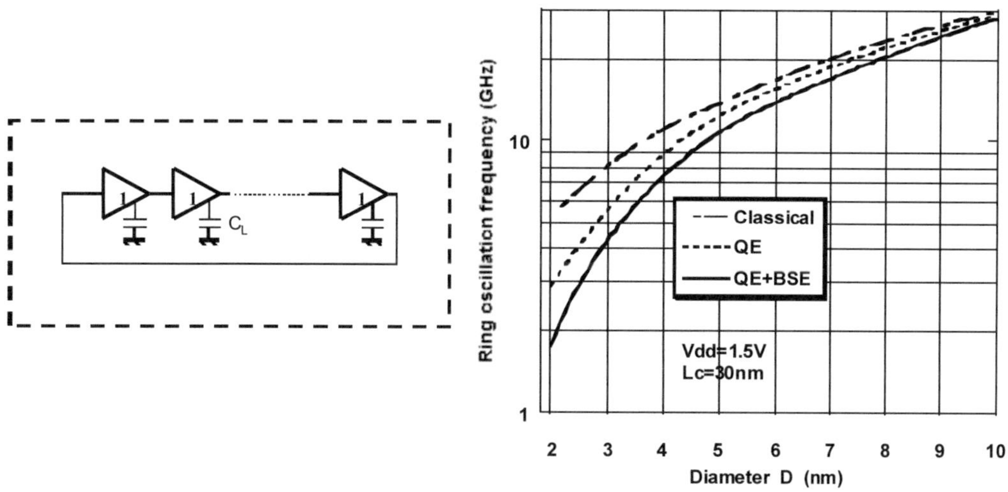

Fig. 15. Impact of BSE on the ring oscillator frequency versus the nanowire diameter. Comparison between classical, quantum (QE) and low dimensions effects (QE+BSE).

5. Conclusion

In this chapter, the potential of silicon nanowires for microelectronics applications was evaluated. Regarding the evolution of transistor architectures, they appear as the best configuration for the gate control of the device. The particular shape with a surrounding gate provides an ideal electrostatic control to immunize transistors against perturbations generated by the scaling down of dimensions. In this work, we have developed a complete model of the electrostatic of transistors based on nanowires. The physical phenomena affecting the electrostatics was considered: short-channel effects due to the channel length reduction or the quantum mechanical effects due to the diameter thinning. Moreover, ultimate mechanisms as the modification of the band shape of silicon material is studied based on advanced simulations (essentially tight-binding Schrodinger-Poisson solving). All this physics (thanks to analytical model development) is transposed to higher simulation levels as characteristics of transistor or small-circuit performances. Following this idea, we have seen the impact of short-channel, quantum or band structure effect on the threshold voltage. For ultra-thin nanowires, we highlighted the necessity to consider all these phenomena to be as close as possible to experimental data. Then, a study of their impact on transport was performed with the analysis of ballistic drain current of single transistor and performances of inverters or ring oscillators. In all cases, the evaluation of performances is inaccurate if quantum or band structure effects are not considered. For example, one decade and a half of difference on OFF-state current or a reduction of factor 2 or 3 on the oscillation frequency show the importance of the electrostatics (and so a realistic modeling) if we envisage nanowires as the future technological solution in microelectronics.

6. References

ATLAS. (2010). Users Manual, SILVACO, 2010.

Autran, J.L.; Munteanu, D.; Tintori, O.; Harrison, S.; Decarre, E. & Skotnickin, T. (2004). Quantum-mechanical analytical modeling of threshold voltage in long-channel double-gate MOSFET with symmetric and asymmetric gates, *Proceedings of 7th Int. Conf. Modeling and Simulation of Microsystems, MSM'2004*, pp. 163-166, ISBN 0-9728422-8-4, Boston, Massachusets, USA, March 07-11, 2004

Autran, J.L.; Nehari, K. & Munteanu, D. (2005). Compact modeling of the threshold voltage in silicon nanowire MOSFET including 2D-quantum confinement effects, *Molecular Simulation*, Vol. 31, No. 12, (October 2005), pp. 839-843, 0892-7022

Baccarani, G. (2008). Device Physics for the Modeling of Future Nanoscale Silicon Devices, *Proceedings of SINANO Summer School on Device Modeling*, Bertinoro, Italy, September 1-5, 2008

Banna, S.R.; Chan, P.C.H.; Ko, P.K.; Nguyen, C.T. & Chan, M. (1995). Threshold voltage model for deep-submicrometer fully depleted SOI MOSFETs, *IEEE Trans. Electron Devices*, Vol. 11, No. 5, (April 1995), pp. 487-495, 0026-2714

Barral, V.; Poiroux, T.; Andrieu, F.; Faynot, O. ; Ernst, T.; Brevard, L.; Fenouillet-Beranger, C.; Lafond, D. ; Hartmann, J.M.; Vidal, V.; Allain, F.; Daval, N.; Cayrefourcq, I.; Tosti, L. ; Munteanu, D.; Autran, J.L. & Deleonibus, S. (2007). Strained FDSOI CMOS technology scalability down to 2.5nm film thickness and 18nm gate length with a TiN/HfO2 gate stack, *Proceedings of IEDM Tech. Dig.*, pp. 61-64, ISBN 978-1-4244-1507-6, Washington, USA, December 10-12, 2007

Barral, V.; Poiroux, T.; Vinet, M.; Widiez, J. ; Previtali, B.; Grosgeorges, P.; Le Carval, G.; Barraud, S.; Autran, J.L.; Munteanu, D. & Deleonibus, S. (2007). Experimental determination of the channel backscattering coefficient on 10-70 nm-metal-gate, Double-Gate transistors, , *Solid State Electron.*, Vol. 51, No. 4,(April 2007), pp. 537-542, 0038-1101

Barraud, S.; Bonno, O. & Cassé, M. (2009). Phase-coherent quantum transport in silicon nanowires based on Wigner transport equation: Comparison with the nonequilibrium-Green-function formalism, *Journal of Applied Physics.*, Vol. 106, No. 6, (September 2009) 0021-8979

Collinge, J.P. (2007). Multiple-gate SOI MOSFETs, *Solid State Electron.*, Vol. 48, No. 9, (September 2007), pp. 2071-2077, 0038-1101

Dupre, C.; Hubert, A.; Becu, S.; Jublot, M.; Maffim-Alvaro, V. & Vizioz, C. (2008). 15nm-diameter 3D stacked nanowires with independent gates operation: ΦFET, *Proceedings of IEDM Tech. Dig.*, pp. 749-752, ISBN 978-1-4244-2377-4, San Francisco, California, USA, December 15-17, 2008

Dura, J.; Martinie, S.; Munteanu, D.; Jaud, M.A.; Barraud, S. & Autran, J.L. (2010). Analytical model of quantum threshold voltage in short-channel nanowire MOSFET including band structure effects, *Proceedings of NSTI-Nanotech WCM*, Vol. 2, pp.801-804, ISBN 978-1-4398-3402-2, Anaheim, California, USA, June 21-24, 2010

Dura, J.; Martinie, S. ; Munteanu, D. ; Triozon, F. ; Barraud, S. ; Niquet, Y.M. ; Barbé, J.C. & Autran, J.L. (2011). Analytical model of ballistic current for GAA nanowire MOSFET including band structure effects: Application to ring oscillator, *Proceedings of 2011 Ultimate Integration of Silicon*, ISBN 978-1-4577-0089-7, Cork, Ireland, March 14-16, 2011

Hwi Cho, K.; Suk, S.D.; Yeoh, Y.Y.; Li, M.; Yeo, K.H.; Kim, D.W.; Park, D.; Lee, W.S.; Jung, Y.C.; Hong, B.H. & Sung Woo H. (2007). Temperature-dependent characteristics of cylindrical gate-all-around twin silicon nanowire MOSFETs (TSNWFETs). *IEEE Electron Devices Letters*, Vol. 28, No. 12, (December 2007), pp. 1129-1131, 0741-3106

ITRS. (2009). In: *International Technology Roadmap for Semiconductors*, 05.08.2011, *Available from:* http://www.itrs.net/Links/2009ITRS/Home2009.htm

Jimenez, D.; Iniguez, B.; Sune, J.; Marsal, L.F.; Pallares, J.; Roig, J. & Flores, D. (2004). Continuous analytic I-V model for surrounding-gate MOSFETs, *IEEE Electron Device Lett.*, Vol. 25, No. 8, (August 2004), pp. 571-573, 0741-3106

McKelevey, J.P.; Longini, R.L. & Brody, T.R. (1961). Alternative approach to the solution of added carrier transport problems in semiconductors, *Physical Review*, Vol. 123, No. 1, (July 1961), pp. 2736-2743, 0010.1103

Munteanu, D. & Autran, J.L. (2003). Two-dimensional Modeling of Quantum Ballistic Transport in Ultimate Double-Gate SOI Devices, *Solid State Electron.*, Vol. 47, No. 7, (July 2003), pp. 1219-1225, 0038-1101

Munteanu, D.; Autran, J.L.; Harrison, S.; Nehari, K.; Tintori, O. & Skotnicki, T. (2005). Compact model of the quantum short-channel threshold voltage in symmetric Double-Gate MOSFET, *Molecular Simulation*, Vol. 31, No. 12, (July 2005), pp. 1911-1918, 0022-3093

Nehari, K.; Cavassilas, N.; Autran, J.L.; Bescond, M.; Munteanu, D. & Lannoo, M. (2006). Influence of band structure on electron ballistic transport in silicon nanowire MOSFET's: An atomistic study, *Solid State Electron.*, Vol. 50, No. 4, (April 2006), pp. 716-721, 0038-1101

Neophytou, N.; Klimeck, G. & Lundstrom, M.S. (2008). Bandstructure effects in silicon nanowire electron transport, *IEEE Trans. Electron Devices*, Vol. 43, No. 5, (June 2008), pp. 732-738, 0018-9383

Niquet, Y.M.; Delerue, C.; Allan, G. & Lannoo, M. (2000). Method for tight-binding parametrization: Application to silicon nanostructures, *Physical Review B*, Vol. 62, No. 8, (August 2000), pp. 5109-5116, 0163-1829

Niquet, Y.M.; Lherbier, A.; Quang, N.H.; Fernández-Serra, M.V.; Blase, X. & Delerue, C. (2006). Electronic structure of semiconductor nanowires, *Physical Review B*, Vol. 73, No. 16, (April 2006), pp. 5319-5332, 1098-0121

Sarrazin, E.; Barraud, S.; Triozon, F. & Bournel, A. (2008). A self-consistent calculation of band structure in silicon nanowires using a Tight-Binding model, *Proceedings of SISPAD*, No. 5, pp. 349-352, ISBN 978-1-4244-1753-7, Hakone, Japan, September 9-11, 2008

Suk, S.D.; Lee, S.Y.; Kim, S.M.; Yoon, E.J. & Kim, M.S. (2007). Investigation of nanowire size dependancy on TSNWFET, *Proceedings of IEDM Tech. Dig.*, pp. 891-894, ISBN 978-1-4244-1507-6, Washington, USA, December 10-12, 2007

Suzuki, K.; Tosaka, Y. & Sugii, T. (1996). Analytical threshold voltage for short channel n+-p+ double-gate SOI MOSFETs, *IEEE Trans. Electron Devices*, Vol. 43, No. 5, (July 1996), pp. 1166-1168, 0018-9383

Tachi, K.; Casse, M.; Jang, D.; Dupre, C.; Hubert, A.; Vulliet, N.; Maffini-Alvaro, V. ; Vizioz, C.; Carabasse, C.; Delaye, V.; Hartmann, J.M.; Ghibaudo, G.; Iwai, H.; Cristoloveanu, S.; Faynot, O. & Ernst, T. (2009). Relationship between mobility and high-k interface properties in advanced Si and SiGe nanowires, *Proceedings of IEDM Tech. Dig.*, pp. 313-316, ISBN 97-4244-5640-6, Baltimore, USA, December 7-9, 2009

TBSIM. (2011). In: *TB-SIM homepage*, 04.10.2011, *Available from:* http://inac.cea.fr/L_Sim/TB_Sim/index.html

Yu, B.; Lu, H.; Liu, M. & Taur, Y. (2007). Explicit continuous models for double-gate and surrounding-gate MOSFETs, IEEE Trans. Electron Devices, Vol. 54, No. 10, (October 2007), pp. 2715-2722, 0018-9383

Zervos, M. & Feiner, L.F. (2004). Electronic structure of piezoelectric double-barrier InAs/InP/InAs/InP/InAs (111) nanowires, *Journal of Appl. Phys.*, Vol. 95, No. 1, (January 2004), pp. 281-291, 0021-8979

Part 6

Electrostatic Actuation

6

New Approach to Pull-In Limit and Position Control of Electrostatic Cantilever Within the Pull-In Limit

Ali Yildiz[1], Cevher Ak[1] and Hüseyin Canbolat[2]
[1]*Mersin University, Electrical and Electronics Engineering Department, Mersin,*
[2]*Yildirim Beyazit University, Department of Electronics and Communication Engineering, Ankara,*
[1,2]*Turkey*

1. Introduction

Since electrostatic cantilevers are very easy to fabricate, have small dimension, and consume low power, they have been very popular as a sensor. They had been used as a capacitive pressure sensor for measuring blood pressure (Hin-Leung Chau & Wise, 1988), as a microwave switch (Dooyoung Hah et al., 2000), as an air flow sensor (Yu-Hsiang Wang et al., 2007), as a micro-actuator for probe-based data storage (Lu & Fedder, 2004), and in well known commercial applications like inkjet head (Kamusuki et al., 2000), and optical scanners (Schenk et al., 2000).

An electrostatic MEMS cantilever is a simple capacitor consists of two parallel conductive plates. The bottom conductive plate is coated on a substrate and fixed on it, the top plate is suspended with a surface area A. The top electrode is separated by a gap spacing d above the bottom one and fixed from one end. The other end is free to move. When a potential difference (V) is applied between electrodes, free end will tilt downwards (δ) due to electrostatic force (Fig.1.)

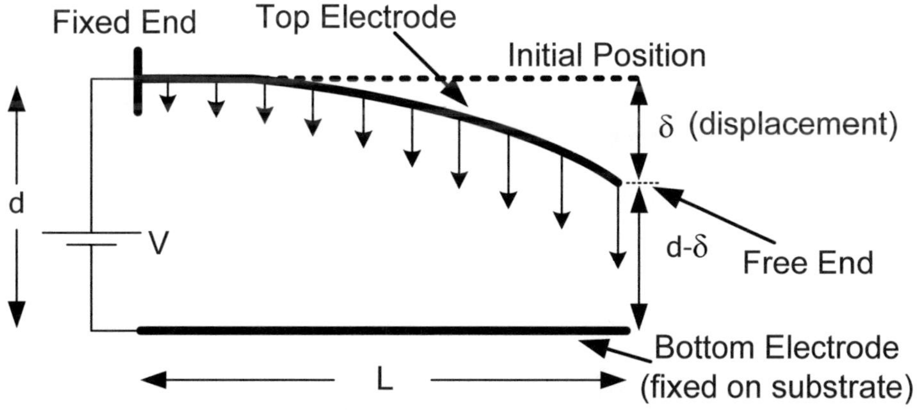

Fig. 1. Electrostatic actuator (Side View)

When potential difference is removed, top electrode will come back to its initial (original) position due to restoring force of the bended structure. It is also sometimes called as spring force. If applied potential is increased beyond the some limit, top electrode will collapse onto the bottom electrode. While spring force term is proportional with the displacement, electrostatic force is proportional to square of the displacement. Hence, after some point, spring force cannot balance the electrostatic force any more. Then, top electrode collapses onto the bottom electrode. This point is named as pull-in limit.

Because cantilever is an electromechanical coupled system, its behavior is non-linear. Thus, having an analytical formula for pull-in limit is impossible. So, there is no simple formula to calculate it. People have been using lumped model (Hyun-Ho Yang et al., 2010; Seeger & Boser, 1999; Nielson & Barbastathis, 2006; Faris et al., 2006; Mol et al., 2007; Chowdhury et al., 2006; Owusu & Lewis, 2007) for decades. Lump model estimates the pull-in limit as one-third of the initial gap (d/3). However, experimental part of a study showed that pull-in limit is at a different value (Hu et al., 2004). Hu et al. utilized a linearized governing equation of a cantilever to demonstrate their analytical approach. They obtained the total energy expressions including the kinetic energy, strain energy, and electric potential energy. The total energy expression was substituted into Hamilton's principle, and obtained partial differential equation with a nonlinear force term. This force term expanded by Taylor series about the equilibrium position and higher order terms neglected with an assumption of small displacement. At the end, structure-electrostatic coupling linear partial differential equation was obtained. Since small deflection was assumed and Taylor series expanded about equilibrium position, as the cantilever tip gets away from initial (original) position, error percentage also gets bigger (as high as 10%) when the tip is closer to pull-in limit(Hu et al., 2004). A later study has used Generalized Differential Quadrature Method which is accurate and efficient way to analyze a linear vector space by high-order polynomial approximation (Sadeghian et al., 2007). This approach gives smaller error in some measurements but not in all of them. Error gets as high as 5% when we close to pull-in limit.

When it is also checked by a software (ANSYS) which utilizes finite element method, pull-in limit seems to be at around 44% of initial gap which is consistent with experimental results (Hu et al., 2004; Sadeghian et al., 2007). Table 1 shows some simulation results for different initial gaps.

Initial Gap(μm)	**Pull-in Gap(μm)**	**Pull-in Gap/Initial Gap**
2	0.881	0.4405
5	2.203	0.4406
10	4.403	0.4403

Table 1. Ansys simulation Pull-in results of a cantilever with L = 200 μm.

These results show that the lumped model is not a very good approximation of the system. In fact, the cantilever beam system has two constrains: fixed end of the top electrode has zero displacement and zero angle even when voltage applied between electrodes. However, lumped model considers only second constrain for the sake of simplicity.

2. Lumped model

As it can be seen from Fig. 2, only zero angle constrain is considered and zero displacement is ignored. Nevertheless, the model is very simple. Therefore, calculations are easy and pull-in limit can be computed in few steps as one-third of the initial gap.

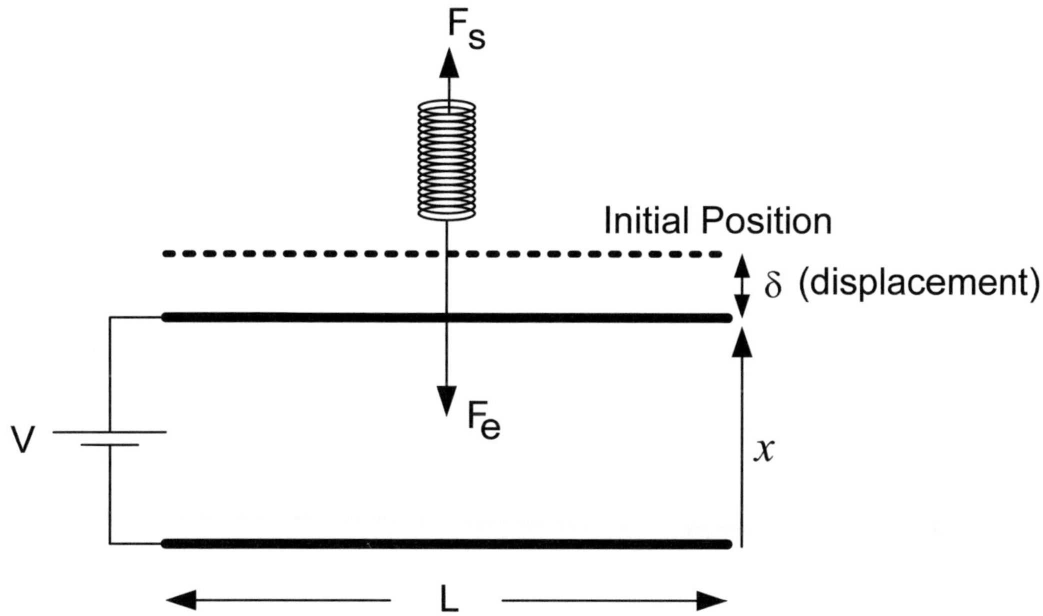

Fig. 2. Lumped model of a cantilever actuator.

Stored energy in a parallel plate capacitor is

$$U = \frac{1}{2} C V^2 \tag{1}$$

Force due to this energy is

$$F = \frac{dU}{dx} \tag{2}$$

Therefore,

$$F = \frac{d}{dx}\left(\frac{1}{2} C V^2\right) = \frac{1}{2}\frac{dC}{dx} V^2 \qquad \left(\frac{dV}{dx} = 0 \text{ since V is constant}\right) \tag{3}$$

Value of a parallel plate capacitor is

$$C = \frac{\varepsilon_0 A}{x} \tag{4}$$

ε_0 is permittivity of free space, A is area of one of the parallel plates, and x is the distance between plates.

So,

$$\frac{dC}{dx} = -\frac{\varepsilon_0 A}{x^2} \tag{5}$$

Then, electrostatic force term is

$$F_e = -\frac{\varepsilon_0 A V^2}{2x^2} \qquad \text{(- sign shows direction of the force)} \tag{6}$$

Electrostatic force term and spring force term will be equal to each other for equilibrium,

$$F_e = F_s \tag{7}$$

$$\frac{\varepsilon_0 A V^2}{2x^2} = k\delta\,, \quad \delta = (d - x) \tag{8}$$

$$2k\left(x^2 d - x^3\right) = \varepsilon_0 A V^2 \tag{9}$$

Potential can be get as

$$V = \sqrt{\frac{2k}{\varepsilon_0 A}\left(x^2 d - x^3\right)} \tag{10}$$

Top electrode will collapse when

$$\frac{dV}{dx} = 0 \tag{11}$$

So, if derivative is taken of Eq. (10) we have

$$2xd - 3x^2 = 0 \tag{12}$$

Then critical x value can be get as

$$x_{critical} = \frac{2}{3}d \tag{13}$$

Displacement of top electrode at the limit condition can be found as

$$\delta_{critical} = \frac{1}{3}d \tag{14}$$

Thus, critical value of the potential difference can be calculated as

$$V_{critical} = \sqrt{\frac{2k}{\varepsilon_0 A}\left(\frac{2}{3}d\right)^2 d - \left(\frac{2}{3}d\right)^3} \tag{15}$$

Therefore, it can be simplified as

$$V_{critical} = \sqrt{\frac{8}{27}\frac{kd^3}{\varepsilon_0 A}} \tag{16}$$

k can be obtained for a cantilever as (Saha et al., 2006)

$$k = \frac{2}{3}Ew\left(\frac{t}{L}\right)^3 \tag{17}$$

and insert this in Eq. (16), and replace *A* with w*L*. We can get

$$V_{critical} = \sqrt{\frac{16}{81}\frac{Ed^3t^3}{\varepsilon_0 L^4}} \tag{18}$$

3. Bisection model

Since model has two different sections, it is named as *Bisection Model* and can be seen in Fig.3. Bisection model considers two constrains of fixed end of the top electrode. Fixed end has both zero displacement and zero angle while the other end is free to tilt linearly around pivot. Pivot point is placed 1/3 of the cantilever length from fixed end of top electrode. There is no specific reason for 1/3 ratio exactly. Since, the left side of the cantilever is fixed, there is very small movement at the left side. So, we model the structure in a that way 1/3 of the cantilever is not moving at all and rest is moving linearly around the break point (pivot). By doing this, we still have very simple model as Lumped model and also consider both constraint of fixed end. Therefore, we can have a simple formula for the structure. Electrostatic force is placed at the free end of the top electrode since this end is close to bottom electrode and force gets its biggest value over there. Restoring force is placed at the one third of the movable part since it has to be close to pivot.

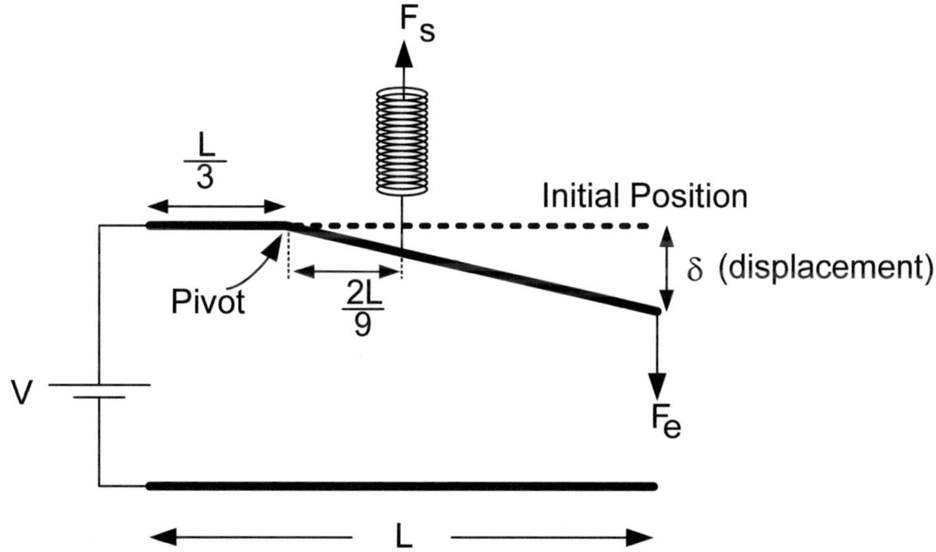

Fig. 3. New Approach (Bisection Model) to cantilever actuator.

In this model, capacitance of the system has two parts (see Fig. 4.)

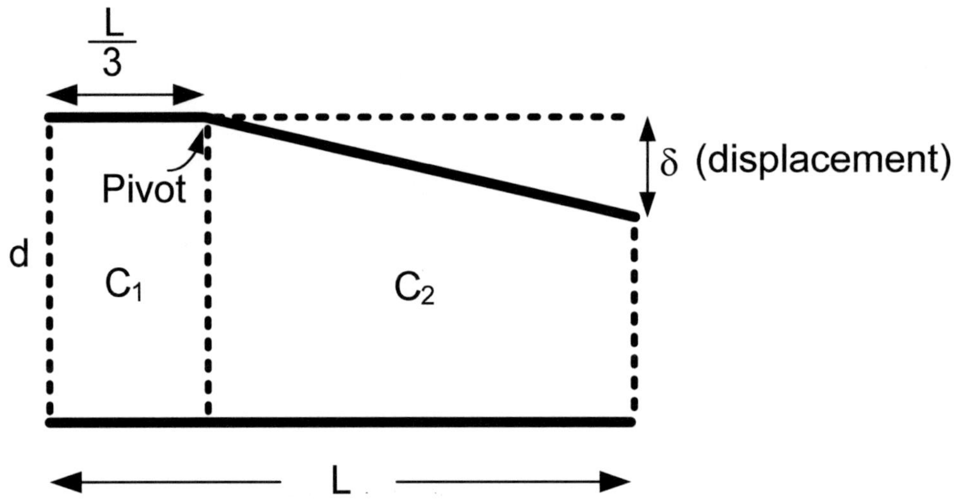

Fig. 4. Bisection Model capacitor calculation.

C_1 and C_2 can be found as

$$C_1 = \varepsilon_0 \frac{wL}{3d} \quad \text{and} \quad C_2 = \frac{2\varepsilon_0 wL}{3\delta} \ln\left(\frac{d}{d-\delta}\right) \tag{19}$$

Where w and L are width and length of the cantilever respectively.

Therefore, total capacitance of the system is

$$C_T = C_1 + C_2 = \frac{1}{3}\varepsilon_0 wL\left(\frac{1}{d}+\frac{2}{\delta}\ln\left(\frac{d}{d-\delta}\right)\right) \tag{20}$$

And electrostatic force term can be obtained as

$$F_e = \frac{1}{2}\left[\left(\frac{d}{d\delta}(C_1+C_2)\right)V^2+\left(\frac{d}{d\delta}V^2\right)(C_1+C_2)\right] \tag{21}$$

Since V (potential difference between electrodes) is constant between plates and C_1 is also constant, equation can be written shortly as

$$F_e = \frac{1}{2}\frac{dC_2}{d\delta}V^2 \tag{22}$$

Since

$$\frac{dC_2}{d(d-\delta)} = \frac{2}{3}\left(-\frac{\varepsilon_0 wL\ln\left(\dfrac{Ld}{Ld-L\delta}\right)}{\delta^2}+\frac{\varepsilon_0 wL^2}{\delta L(d-\delta)}\right) \tag{23}$$

The electrostatic force term can be obtained as

$$F_e = \frac{1}{3}\varepsilon_0 wL\left(\frac{-\ln\left(\frac{d}{d-\delta}\right)d+\ln\left(\frac{d}{d-\delta}\right)\delta+\delta}{\delta^2(d-\delta)}\right)V^2 \tag{24}$$

The storing force term can be written as

$$F_s = k\frac{\delta}{3} \tag{25}$$

and k can be obtained for a cantilever as (Saha et al., 2006)

$$k = \frac{2}{3}Ew\left(\frac{t}{L}\right)^3 \tag{26}$$

Where E and t are Young's modulus and thickness of top electrode.

In the Bisection Model only $2L/3$ part of the upper electrode is free to move. So, by substituting $2L/3$ to L, k can be calculated as

$$k = \frac{9}{4}Ew\left(\frac{t}{L}\right)^3 \tag{27}$$

for Bisection Model. Since moments of the electrostatic and restoring forces are equal, we can write

$$\frac{6}{9}F_e = \frac{2}{9}F_s \tag{28}$$

Therefore, the relation between applied voltage and displacement is given by

$$V = \left(\frac{3}{4}\frac{Et^3\delta^3(d-\delta)}{\varepsilon_0 L^4(-\ln\left(\frac{d}{d-\delta}\right)d+\ln\left(\frac{d}{d-\delta}\right)\delta+\delta}\right)^{\frac{1}{2}} \tag{29}$$

This equation is valid not only for pull-in limit, but also for all values within the pull-in limit.

To find the pull-in limit, we have to take derivative of V with respect to δ. So;

$$\frac{dV}{d\delta} = \frac{-\frac{\sqrt{3}Et^3}{4}\left(4\delta^2-3d\delta+\left(3\delta^2-6\delta d+3d^2\right)\ln\left(\frac{d}{d-\delta}\right)\right)}{\left(\delta Et^3(d-\delta)L^4\varepsilon_0\left(\ln\left(\frac{d}{d-\delta}\right)(\delta-d)+\delta\right)^3\right)^{\frac{1}{2}}} = 0 \tag{30}$$

There is no analytical solution to this equation. Hence, computational result has been obtained as:

$$\frac{\delta_{max}}{d} = 0.440423 \tag{31}$$

In addition, when a graph of the equation is drawn, it shows stable and unstable regions (Fig. 5). It can be seen that the Pull-in Limit is at the 44% of the initial gap.

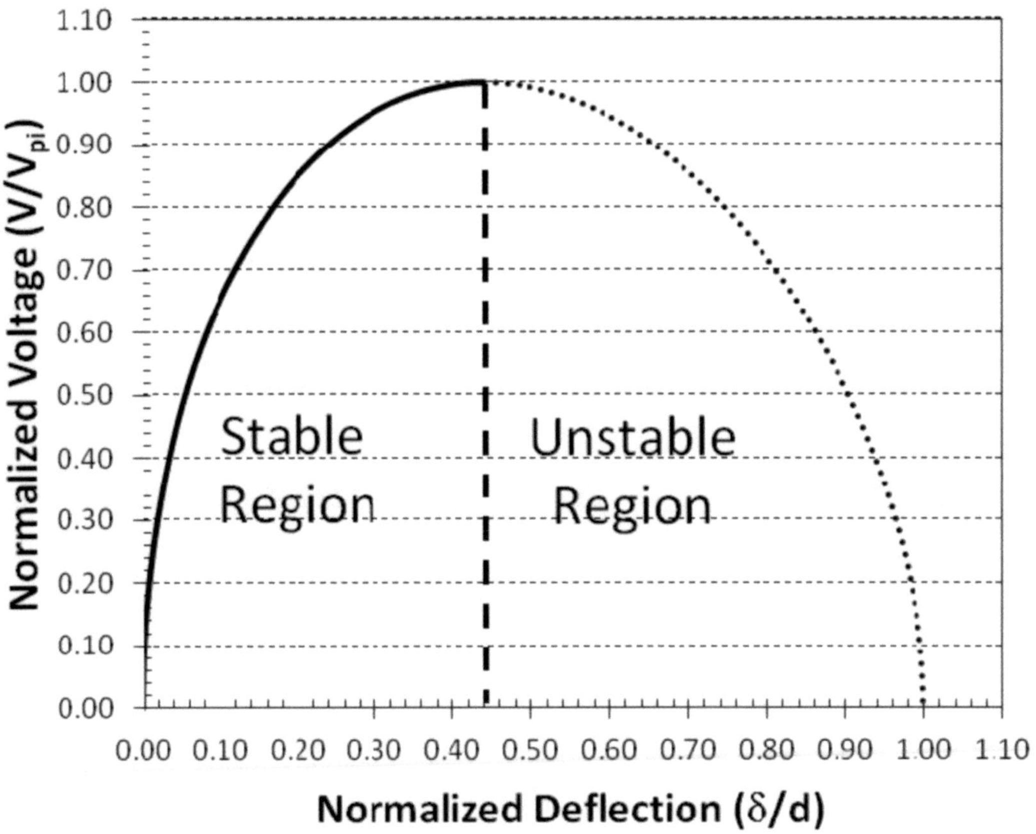

Fig. 5. Stable and Unstable Regions of Bisection Model.

Thus, critical value of the potential difference can be calculated as

$$V_{critical} = \sqrt{\frac{69}{500}\frac{kd^3}{\varepsilon_0 A}} \tag{32}$$

and insert Eq. (27) into Eq. (32), and replace *A* with w*L*. We can get

$$V_{critical} = \sqrt{\frac{621}{2000}\frac{Ed^3t^3}{\varepsilon_0 L^4}} \tag{33}$$

Eq. (33) is very similar with Lumped Model's Eq. (18) except coefficient which helps to get better results when compared with Lumped Model.

Table 2 and 3 show comparison of Ansys simulation results and values obtained from Bisection Model and percentage error for cantilever with different lengths.

(Initial gap=2μm)	**V_{max}(V) From Bisection Model**	**V_{max}(V) From ANSYS**	**% Error**
L=150	27.4707	27.3408	0.475
L=200	15.4523	15.4179	0.223
L=250	9.8894	9.8985	0.092
L=300	6.8677	6.8284	0.575
L=400	3.8631	3.8604	0.069
L=500	2.4724	2.4715	0.036

Table 2. Comparison of V_{max} (Pull-in Voltage) values for cantilevers with different lengths

(L=150μm, d=2μm)	**Voltage (V) From Bisection Model**	**Voltage (V) From ANSYS**	**% Error**
δ= 0.01415 (0.71 %)	5.0549	5.0	1.098
δ= 0.05829 (2.91%)	10.1072	10.0	1.072
δ= 0.1386 (6.93%)	15.1544	15.0	1.030
δ= 0.2714 (13.57%)	20.1942	20.0	0.971
δ= 0.5165 (25.83%)	25.2056	25.0	0.822
δ= 0.6028 (30.14%)	26.1918	26.0	0.737
δ= 0.7419 (37.10%)	27.1588	27.0	0.588
δ= 0.7654 (38.27%)	27.2542	27.1	0.569
δ= 0.7963 (39.82%)	27.3518	27.2	0.558
δ= 0.8146 (40.73%)	27.3949	27.25	0.531
δ= 0.8808 (44.04%)	27.4628	27.34	0.449

Table 3. Comparison of Voltage values for arbitrary δ displacements.

Table 4 shows comparison of previous experimental, analytical results (Hu et al., 2004; Sadeghian et al., 2007), Bisection Model Result, and percentage error with respect to experimental results. Ansys simulation results also added for comparison.

Voltage(V)	Experimental (Hu et al., 2004)	Analytical (Hu et al., 2004) / (Error)	Analytical (Sadeghian et al., 2007) / (Error)	Bisection Model / (%Error)	ANSYS
20	90.5	90.2 / (0.3%)	90.2 / (0.3%)	90.4 / (0.1%)	90.4
40	84.6	84.3 / (0.4%)	84.1 / (0.6%)	85.1 / (0.6%)	85.1
60	70.0	71.5 / (2.1%)	69.1 / (1.2%)	73.2 / (4.5%)	73.2
65	64.0	67.2 / (5.0%)	59.6 / (6.9%)	67.4 / (5.3%)	67.6
67	59.0	65.0 / (10.2%)	-	64.1 / (8.7%)	64.5

Table 4. Comparison of displacements for different voltage values. Errors are respect to experimental results.

4. Conclusion

Values calculated from Bisection Model are very close to those obtained from ANSYS. Especially, when the displacement is larger than 10% of the initial gap, all the errors are within 1% (see Table 3). Therefore, Bisection Model not only gives a better pull-in limit when compared with previous lumped model, but also has simpler analytical result when compared with previous discrete models. At the same time, it gives satisfactory results for applied voltages for given displacements of top electrode's free end. Bisection Model is also very successful when compared to experimental studies. Percentage error level of Bisection Model is comparable when displacement is small, and gets better when displacement is close to pull-in limit (see Table 4). Bisection Model also gives a simple formula to use instead of using numerical methods which is time consuming and requires computation capacity.

5. References

Chowdhury, Sazzadur.; Ahmadi, Majid. & Miller, W. C. (2006). Pull-in Voltage Study of Electrostatically Actuated Fixed-Fixed Beams Using a VLSI On-Chip Interconnect Capacitance Model. *IEEE journal ofMEMS,* vol.15, no.3, pp.639-651, ISSN 1057-7157

Dooyoung Hah.; Euisik Yoon. & Songheol Hong. (2000). A Low-Voltage Actuated Micromachined Microwave Switch Using Torsion Springs ans Leverage. Microwave Symposium Digest, vol.1, pp.157-160, ISSN 0149-645X

Faris, W.F.; Mohammed H.M.; Abdalla, M.M. & Ling, C.H. (2006). Influence of Micro-Cantilever Geometry and Gap on Pull-in Voltage. *Dans Symposium on Design, Test,*

Integration and Packaging of MEMS/MOEMS, DTIP 2006, Stresa, Lago Maggiore : Italie (2006)

Hin-Leung Chau. & Wise, K.D. (1988). An ultraminiature solid-state pressure sensor for a cardiovascular catheter. *IEEE Trans. Electronic Devices*, vol.35, No.12, pp. 2355-2362, ISSN 0018-9383

Hu, Y.C.; Chang, C.M. & Huang S.C. (2004). Some Design Considerations on the electrostatically actuated microstructures. *Sensors and Actuators,* vol.112, no.11, pp.155-161 ISSN 0924-4247

Hyun-Ho Yang.; Jeong Oen Lee. & Jun-Bo Yoon. (2010). Maneuvering Pull-in Voltage of an Electrostatic Micro-Switch by Introducing a Pre-charged Electrode. *MEMS, 2010 IEEE 23rd International Conference on,* Wanchai, Hong Kong, 24-28 Jan., pp.747-750, ISSN 1084-6999

Kamusuki, S.; Fujii, M.; Takekoshi, T. & Tekuza, C. (2000) A high resolution, electrostatically driven commercial inkjet head. *Proceedings of IEEE MEMS 2000 conference,* Miyazaki, Japan, 23-27 Jan., pp. 793-798.

Lu, M.S.-C. & Fedder, G.K. (2004). Position Control of Parallel-Plate Microactuators for Probe-Based Data Storage. *IEEE Journal of MEMS,* vol.12 no.5, pp.759-769, ISSN 1057-7157

Mol, Lukas.; Rocha, Luis A.; Cretu, Edmond. & Wolffenbuttel, Reinoud F. (2007). Full-Gap Positioning of Parallel-Plate Electrostatic MEMS Using On-off Control. IEEE Int. Symposium on Industrial Electronics, ISIE2007, Vigo, Spain, pp.1464-1468

Nielson G. N. & Barbastathis, George. (2006). Dynamic Pull-in of Parallel-Plate and Torsional Electrostatic MEMS Actuators. *IEEE MEMS Journal,* vol.15, no.4, pp.811-821, ISSN 1057-7157

Owusu, Kwadwo O. & Lewis, Frank L. (2007). Solving the Pull-in Instability Problem of Electrostatic Microactuators Using Nonlinear Control Techniques. *Nano/Micro Engineered and Molecular Systems, 2007. NEMS '07. 2nd IEEE International Conference on,* 16-19 Jan, Bangkok, pp.1190-1195

Sadeghian, Hamed.; Rezazadeh, Ghader. & Osterberg, Peter M. (2007). Application of the Generalized Differential Quadrature Method to the Study of Pull-in Phenomena of MEMs Switches. *Journal of Microelectromechanical Systems,* Vol. 16,No. 6, pp. 1334-1340 ISSN 1057-7157

Saha, S. C.; Hanke, U.; Jensen, G. U. & Saether, T. (2006). Modeling of Spring Constant and Pull-down Voltage of Non uniform RF MEMS Cantilever. *IEEE Behavioral Modeling and Simulation Workshop,* s.56-60,

Schenk, H.; Durr, P.; Kunze, D.; Lakher, H. & Kuck, H. (2000). An electrostatically excited 2D-micro scanning-mirror with an in plane configuration of the driving electrodes. *Proceedings of IEEE MEMS 2000 Conference,* Miyazaki, Japan, 23-27 Jan., pp. 473-478.

Seeger, Joseph I. & Boser, Bernhard E. (1999). Dynamics and Control of Parallel-Plate Actuators Beyond the Electrostatic Instability. *10th International Conf. on Solid-Stade Sensor and Actuators,* Sendai, Japan, 7-9 June, pp.474-477

Yu-Hsiang Wang.; Chia-Yen Lee. & Che-Ming Chiang. (2007). A Mems-based Air Flow Sensor with a Free-Standing Microcantilever Structure. *Sensors,* vol.7, no.10, pp.2389-2401